D0174171

The Elements of Technical Writing

Thomas E. Pearsall

University of Minnesota, Emeritus

Allyn and Bacon

Boston   London   Toronto   Sydney   Tokyo   Singapore

For Anne,
my wife and best friend

Vice President: Eben Ludlow
Editorial Assistant: Liz Egan
Marketing Manager: Lisa Kimball
Production Administrator: Donna Simons
Editorial-Production Service: Susan Freese, Communicáto, Ltd.
Design and Electronic Composition: Denise Hoffman, Glenview Studios
Composition/Prepress Buyer: Linda Cox
Manufacturing Buyer: Megan Cochran
Cover Administrator: Suzanne Harbison

Copyright © 1997 by Allyn & Bacon
A Viacom Company
160 Gould Street
Needham Heights, MA 02194

Internet: www.abacon.com
America Online: keyword:College Online

All rights reserved. No part of the material protected by this copyright notice
may be reproduced or utilized in any form or by any means, electronic or
mechanical, including photocopying, recording, or by any information storage
and retrieval system, without the written permission of the copyright owner.

Library of Congress Cataloging-in-Publication Data

Pearsall, Thomas E.
    The elements of technical writing / Thomas E. Pearsall.
        p.   cm.
    Includes bibliographical references and index.
    ISBN 0-205-18895-8 (alk. paper)
    1. Technical writing.   I. Title.
T11.P392   1997
808'.0666—dc20   96-27196
                                                  CIP

Printed in the United States of America

10 9 8 7 6 5 4 3 2 1   01 00 99 98 97 96

Contents

5   Use Good Page Design   32

6   Think Visually   47

7   Write Ethically   66

Part Two
The Formats of Technical Writing   *75*

8   Purposes of Formats   77

9   Elements of Reports   81

10 Formats of Reports   106

Preface

In the 30 years that technical writing has grown into an important academic discipline, it has also grown in size and complexity. Technical writing textbooks, including my own, have grown in size and complexity with the discipline.

Eben Ludlow, my editor at Allyn and Bacon, suggested that there might be readers who needed a concise introduction to technical writing, one that concentrated on the essential elements of the discipline. The challenge was an interesting one: Tell readers what they need to know to turn out useful reports and correspondence and nothing more. I have enjoyed pursuing that challenge.

Technical writing, purely and simply, is writing about subjects in technical disciplines. Whether agriculture, economics, engineering, or zoology, a technical discipline will have its *technics*—that is, its theories, principles, arts, and skills. It will generate reports and correspondence. The hallmark of such reports and correspondence is that they present objective data to *convince,* rather than emotion to *persuade.*

People in every technical discipline will propose research, report progress on research, and report the results of the research. In addition, people will instruct, transmit information, and argue for the validity of their opinions. Such is the stuff of technical writing, and such is the stuff of this book.

Plan of the Book

After paring down to the essentials, I was left with seven chapters on the principles of technical writing and four on the formats—eleven chapters in all. Fearful that readers might need a little more help, I added four appendixes showing sample reports as a way of tying things together.

The seven chapters in Part One present the essential principles of good technical writing:

- Chapter 1, Know Your Purpose
- Chapter 2, Know Your Audience
- Chapter 3, Organize Your Content around Your Purpose and Audience
- Chapter 4, Write Clearly and Precisely
- Chapter 5, Use Good Page Design
- Chapter 6, Think Visually
- Chapter 7, Write Ethically

Part Two covers some of the structural and organizational features of technical writing:

- Chapter 8, Purposes of Formats, explains the importance of formats in providing accessibility, selective reading, and comprehension for readers.
- Chapter 9, Elements of Reports, describes such elements as title pages, abstracts, introductions, and documentation.
- Chapter 10, Formats of Reports, provides formats for major reports, such as research reports, recommendation reports, and proposals.
- Chapter 11, Formats of Correspondence, shows how to format letters and memorandums and how to write letters ofapplication and résumés.

The four appendixes provide illustrative samples of the following:

- A letter analytical report
- A student proposal
- A progress report
- An empirical research report

Acknowledgments

I have, in many places, acknowledged the sources of ideas and materials. Most often, the ideas have come from a combination of places that would be impossible to document. They are the materials that those of us in technical writing have freely shared with each other for years. But I would be remiss if I did not acknowledge here some of the people who have shaped my thoughts the most: Paul Anderson, Virginia Book, W. Earl Britton, David Carson, Mary Coney, Donald Cunningham, Herman Estrin, Jay Gould, John Harris, Ken Houp, Ann Laster, Fred MacIntosh, Victoria Mikelonis, John Mitchell, Nell Ann Pickett, Janice Redish, Tom Sawyer, Jim Souther, John Walter, Tom Warren, Arthur Walzer, and Mike White. Blessings on them all, and double blessings on anyone I have neglected to mention.

I also thank my colleagues who reviewed this book in manuscript for Allyn and Bacon and made many useful suggestions: Dan Jones, University of Central Florida, and Art Walzer, University of Minnesota.

T. E. P.

PART ONE

The Seven Principles of Technical Writing

1 Know Your Purpose

2 Know Your Audience

3 Choose and Organize
   Your Content around Your
   Purpose and Audience

4 Write Clearly and Precisely

5 Use Good Page Design

6 Think Visually

7 Write Ethically

1

Know Your Purpose

What goal do you have in writing? What objective do you wish to achieve? What result do you hope for? What do you want to accomplish? In short, all these questions address what you see as your *purpose* in whatever it is you are writing, whether a letter, a résumé, or a report.

Basically, *technical writing* has only two purposes: (1) to inform and (2) to argue. Given that, consider these two examples of purpose:

> To inform technical writers about the various computer programs available for use in editing manuscripts.

> To inform technical writers about the various computer programs available for use in editing manuscripts and to argue for the superiority of several programs over others.

The first purpose—to inform—will be met by writing a report that objectively describes the programs available. The report that satisfies the second purpose—to argue—should be no less objective but will contain an argument that evaluates the programs to demonstrate the superiority claimed, perhaps through a comparison and contrast of their features and simplicity of use. (See Chapter 10 to review the various formats used in writing reports that inform and argue.)

Many variations on reports that inform and argue exist in technical writing. For instance, definitions and descriptions are largely informative. So are written instructions, for the most part, although argument may play a role, as in the following example:

> This new way of doing this procedure is better than that old way.

But much of technical writing is argumentative. Research reports inform readers of the facts that have been uncovered and then argue that the conclusions drawn from those facts are reliable. Similarly, recommendation reports inform, analyze, evaluate, and draw conclusions that support their recommendations; proposals argue for engaging organizations to do certain tasks; and so on.

To be a successful writer, you must have a clear vision of what purpose you have for composing a given piece of writing. And you will be more likely to achieve your purpose if you state it in writing. Here are some examples:

> I want to convince the executives of Elmont Bank to accept my company's proposal to update the software used by their tellers in reporting bank transactions.

> I want to demonstrate to the upper management of Dickens's Sporting Goods Company the feasibility of manufacturing the heads of golf clubs from lightweight titanium.

> My purpose is to instruct the tellers of Elmont Bank how to use the new software we have installed for their use in reporting bank transactions.

> My purpose is to explain to the technicians who manufacture golf clubs for Dickens's Sporting Goods Company the techniques required for working with titanium in casting golf club heads.

Notice that all these statements of purpose mention the *audience:* that is, the person or persons to whom the writing is directed. Understanding your audience is as essential to successful writing as understanding your purpose. We will look at the concept of audience more closely in the next chapter, Know Your Audience.

2

Know Your Audience

Good writers are aware of their audiences. As you visualize your readers, consider these things:

- Their concerns and characteristics
- Their levels of education and experience
- Their attitudes toward your purpose and information

Concerns and Characteristics

Figure 2.1 summarizes the concerns and characteristics of five major audiences you may face as a technical writer and indicates some ways to satisfy the needs of each. For example, notice that executives read primarily for the purpose of making decisions. Thus, an executive will be disappointed by a report that does not clearly state the author's conclusions and back up those conclusions with sufficient but succinct information. Notice also that executives read selectively, skimming and scanning materials. Picture an officer in a large company, facing a desk piled high with reports. Looking over these reports, the officer can make various judgments and exercise options accordingly:

> This entire report directly concerns decisions I have to make; I need to read it carefully and evaluate the information and conclusions presented.

Audience	Concerns and Characteristics
Laypersons	• Read for learning and interest • Have more interest in practice than theory • Need help with science and mathematics • Enjoy and learn from human interest • Require background and definitions • Need simplicity • Learn from simple graphics
Executives	• Read to make decisions • Have more interest in practice than theory • Need plain language • Learn from simple graphics • Need information on people, profits, and environment • Expect implications, conclusions, and recommendations expressed clearly • Read selectively—skimming and scanning • Have self-interests as well as corporate interests
Technicians	• Read for how-to information • Expect emphasis on practical matters • May have limitations in mathematics and theory • May expect theory if higher level
Experts	• Read for how and why things work • Need and want theory • Will read selectively • Can handle mathematics and terminology of field • Expect graphics to display results • Need new terms defined • Expect inferences and conclusions to be clearly but cautiously expressed and well supported
Combined	• One person may combine the attributes of several audiences • Readers may consist of representatives of several audiences

FIGURE 2.1   Audience Concerns and Characteristics

Some of this report concerns my department; I better look through it and read the pertinent parts carefully.

This report looks interesting; I better scan it.

This report contains nothing of concern to me; I will initial and pass it on.

Using the information in this book about good page design (see Chapter 5) and formats of technical writing (see Part Two), you can support executives' thinking processes and allow them to read selectively.

Unlike executives, technicians read primarily for how-to information. For example, you operate as a technician when you consult the instruction manual for, say, your personal computer. You go to that manual to learn how to operate the computer or perhaps to solve a particular problem you are having with it. And you expect the information in the manual to be written in language you clearly understand and organized in a format you can easily access.

In similar fashion, experts read professional books, journals, and papers to keep up to date on the latest research and developments in their fields and to guide their work, and laypeople read things that may interest or entertain them or influence their attitudes and decisions. Keeping your readers' concerns and characteristics in mind is essential to satisfying their needs and to accomplishing your own purpose.

Education and Experience

Based on our individual levels of education and experience, we all learn *vocabulary, concepts,* and *techniques* for doing things. For example, civil engineers understand that the term *safety factor* means "the margin by which a machine exceeds its required performance." The safety factor for most passenger elevators is 7.6; that is, they will carry 7.6 times the average weight of the number of passengers they are specified to carry. In addition, engineers understand the concept of *safety factor.* They could build elevators with a safety factor of 30, but the cost would be so excessive that no one could afford it. Therefore, safety factors are based on a balance of safety and cost.

When civil engineers write, they have to know which technical engineering terms and concepts their readers will likely know and not know.

When writing for fellow engineers, they can assume a shared vocabulary and understanding and leave technical terms and concepts undefined and unexplained. When writing for nonengineers, however, they will have to define and explain as necessary, based on the audience's level of education and experience.

Similarly, through education and experience, we learn techniques for doing things. Consider that the technical writers who produced the computer manual you use had to estimate your knowledge of how to use a keyboard and mouse. If they estimated correctly, you will be a happy reader; if their judgment was incorrect, you will be an unhappy reader.

When you write about what you have learned through your own education and experience, you must consider how well your readers will understand the words, concepts, and techniques you write about. Accordingly, when you think your readers might need help with any of these aspects, offer it.

Attitudes toward Purpose and Information

Consider these possible situations:

> You are writing a report for your supervisor that recommends the purchase of a new, more powerful computer for the office in which the two of you work. You know that your supervisor wants to buy the computer and is looking for support to justify the purchase.

Or maybe the reverse is true:

> Your supervisor does not think the new computer is needed. In fact, she considers it a waste of money.

In the first situation, your supervisor will read your report in a happy frame of mind. If she has the sole decision-making authority to buy the computer, she may not even read the entire report carefully. Your supervisor's approach to your report will be different, however, if she needs to convince someone higher up the chain of command to buy the computer. She will still be happy with your report but will perhaps read it with more care to make sure your argument convincingly expresses your purpose: to justify purchase of the new computer.

In the second situation, your supervisor will be at least skeptical of both your purpose and your information; she may even be hostile. As such, she will read carefully, looking for flaws and weaknesses in your argument for buying the computer. It will take a strong, well-supported argument to convince her.

Consider the position of someone writing a technical manual. He must acknowledge that few people approach using technical manuals with joy in their hearts. Instead, they most likely want to use manuals to find what information they need and would like to do so as easily and quickly as possible. Readers with this attitude want to read manuals selectively; that is, they want to read only those parts they need and skip the rest.

Suppose, for example, that you are writing a manual about how to use a software program that calculates and reports federal income tax. What kinds of information will your readers want? All of them will want to know two basic things: (1) how to load the software into their computers and (2) how to complete a standard 1040 tax form.

Beyond these two skills, however, individual readers will have different needs. For instance, some may need to know how to deal with child-support payments, home office expenses, or individual retirement account (IRA) earnings. Others may need to know how to transfer information from other programs into the tax program or vice versa.

Whatever the specific interest, no one wants to read through unneeded information to get to needed information. Being forced to do so will put anyone in a hostile mood! Therefore, in writing your manual, you should ensure that it is organized and designed to allow readers to find and use needed information easily and to skip the rest.

As these examples indicate, people can approach reading technical documents with widely varying attitudes: agreement or hostility, trust or skepticism, passion or indifference, eagerness or reluctance. One attitude, however, is nearly universal: No one wants to linger. In our busy world, readers want to find what they need in a piece of technical writing and move on. As a technical writer, your job is to help them do just that.

3

Choose and Organize Your Content around Your Purpose and Audience

If you follow the advice given in Chapters 1 and 2 about purpose and audience, you will find it easier to choose your content and organize your report. Make sure you clearly understand the following:

- Your purpose
- Your reader's or readers' concerns and characteristics
- Your reader's or readers' education and experience in the subject area
- Your reader's or readers' attitudes toward your purpose

Because good writing is precise, for a lengthy, important piece of writing, it is a good idea to write down this information, something like this:

Purpose: My purpose is to write a brochure for the members of a health maintenance organization (HMO) that explains the role of folic acid in preventing neural tube malformations (anencephaly and spina bifida) and how women of childbearing age can ensure they get sufficient folic acid in their diets to prevent such malformations during pregnancy.

Readers' concerns and characteristics: The audience I wish to reach, either directly or through relatives or friends, is women of childbearing age. Essentially, this will be an audience of laypeople who read for learning and interest. My readers probably have some concerns about good health and healthy babies, but because they don't know the dangers of neural tube malformations, they have no concerns about them. This audience will require background information and definitions along with graphics and an easy-to-read text.

Readers' education and experience in the subject area: Because a wide cross-section of U.S. society belongs to HMOs, my readers likely have an average level of education (certainly no higher than high school and perhaps as low as the eighth grade). This means I should assume that my readers have no clear knowledge of the role of folic acid in the diet or what neural tube birth malformations are.

Readers' attitudes toward my purpose: Because many people do not worry about health problems until they actually appear, I should assume a certain amount of reader indifference toward my purpose.

When you have completed this analysis, you are ready to choose your content and to organize it in the way that best suits your purpose and audience.

Choosing Content

The principle you should follow in choosing content is a simple one: *Choose the level and amount of content that is needed to fulfill your purpose and your readers' needs—but no more than that.*

Even though this principle seems simple, it is not easy to follow. Most writers—particularly those who are experts in the subjects they are writing about—tell readers more than they really want to know. For example, most owners of VCRs do not care to know anything about

VCRs except how to set them and use them. Giving readers additional information is more than a waste of time. It may actually get in the way of the information that they really do need.

On the other hand, experts in a scientific field reading a report of an experiment in their field probably want information that would not concern or interest nonexperts. Likewise, to be credible and convincing in an argument, you have to provide sufficient information to demonstrate that your conclusions are probably correct.

In other words, choosing content requires thought and judgment on your part. It requires putting yourself in your readers' place. Perhaps the best way to do that is to ask the questions that readers might have. Looking at the HMO example again will illustrate the process.

Imagine for a moment that you are one of the readers of the HMO brochure about folic acid and neural tube malformations. You may be a woman of childbearing age, or perhaps you have friends or relatives who fall into that group. What do you want to know? Your list might include these questions:

What are neural tube malformations?

Will folic acid prevent neural tube malformations?

How much folic acid is required?

When do I need to take folic acid?

Why are neural tube malformations dangerous?

Who is in danger?

What foods or food supplements will provide a sufficient supply of folic acid?

You should choose content that will answer such questions for your readers. Choose enough content so that your answers will be credible and convincing, but do not overload your readers with too much technical detail. For instance, to answer the last question on the list, tell your readers that foods containing folic acid include dark-green leafy vegetables, fruits, beans, whole grains, and breakfast cereals. Indicate, perhaps in a table, how much folic acid each of these foods contains. In addition, tell your readers that most all-purpose vitamin pills contain folic acid. Use a similar method in choosing the content needed to answer the other questions.

Organizing Your Content

When *organizing* your content, as when *choosing* it, keep your readers' needs firmly in mind. In the HMO example, the questions raised suggest that the brochure will be issue or topic oriented. That is, it will give information about specific topics within the general subject of folic acid and neural tube malformations. Many informative reports present a major topic divided into several subtopics, like this:

Topic	Exposure to insecticides
Subtopic 1	Exposure through food
Subtopic 2	Exposure through water
Subtopic 3	Exposure through air
and so on	

In the HMO example, the answers to the questions will be the topics and subtopics for the brochure. How to arrange them still needs to be decided. Your audience analysis told you that most of your readers will not know what neural tube birth malformations are. This suggests that you should first define the term. Definitions grow out of the scheme for a logical definition:

term = genus or class + differentia

In other words, begin your definition by identifying what general class or category your subject belongs to and then provide details to fine-tune your description. For example:

Neural tube malformations are serious birth defects that cause disability or death. They are the most common disabling birth defects, affecting between 1 and 2 out of every 1,000 births in the United States.

There are two main kinds of neural tube malformations: anencephaly and spina bifida. A baby with anencephaly does not develop a brain and dies shortly after birth. Spina bifida is a malformation of the spinal column. If the vertebrae (i.e., bones of the spinal column) surrounding the spinal cord do not close properly during the first 28 days of fetal development, the cord or spinal fluid will bulge through, usually in the lower back.[1]

Look at the first sentence, which follows the scheme for a logical definition:

Neural tube malformations *(term)* are serious birth defects *(genus)* that cause disability or death *(differentia)*.

This definition is then extended with additional information, such as statistics and descriptions; graphic illustrations could be added, as well. Remember, however, that you should not add more detail than your purpose and your readers' needs require.

After you have told your readers what neural tube malformations are, what do you do next? Look at your questions again:

What are neural tube malformations?

Will folic acid prevent neural tube malformations?

How much folic acid is required?

When do I need to take folic acid?

Why are neural tube malformations dangerous?

Who is in danger?

What foods or food supplements will provide a sufficient supply of folic acid?

Next, it would seem to be a good idea to emphasize just how dangerous spina bifida is and who is in danger by describing some of the disabilities involved and adding graphic illustrations of some of them. After you have made a convincing case regarding the dangers of spina bifida, you could then answer the questions concerning folic acid as a preventive measure that all women of childbearing age should follow. The questions concerning folic acid would lend themselves to being organized as a series of subtopics under one topic, like this:

Will folic acid prevent neural tube malformations?

When do I need to take folic acid?

How much folic acid is required?

What foods or food supplements will provide a sufficient supply of folic acid?

Laypeople follow topical discussions well if you use the questions that generated the topics as your organizing device and the actual questions as topic headings. Therefore, your final organizational outline for the brochure might look like this:

What are neural tube malformations?

Why are neural tube malformations dangerous?

Will folic acid prevent neural tube malformations?

　When do I need to take folic acid?

　How much folic acid is required?

　What foods or food supplements will provide a sufficient supply of folic acid?

Although every topical organizational plan grows out of answers to questions, you need not use questions as headings in every situation. For example, for a more expert audience, question-type headings might be inappropriate. But for lay audiences, question-type headings often are the best approach.

You have many organizational schemes available. You may use one of them to organize your entire report and others to organize sections or even paragraphs within it. Here, briefly explained, are some of the most common organizational schemes.

Chronological

In a *chronological* scheme, information is organized by time. You should choose this scheme to report a sequence of events or explain the steps in a process. The following example shows a sequence of events in the order in which they happened:

Event 1　　The accident

Event 2　　The investigation

Event 3　　The trial

and so on

Chronology is also a major organizing pattern in process descriptions:

Step 1   Design the product

Step 2   Build the prototype

Step 3   Test the prototype

Classification and Division

In *classification,* you work from the specific to the general, seeking classifications (i.e., categories) for items. For example, you could group the items *English, history,* and *math* into the classification *college subjects.* In *division,* you work from the general to the specific, seeking items for classifications. So for the classification *college subjects,* you could suggest the items *English, history,* and *math.* In both cases, the end result is the same: a classification and set of items that belong in it:

College subjects

   English

   History

   Math

Be sure that every equal classification or division is based on the same principle. That is, in a classification scheme based on *college subjects—English, history, math,* and so on—do not introduce another equal classification based on *year in college*—for example, *freshman.* It would be appropriate, however, to create a *subclassification* based on *year in college:*

College subjects

   English

      Freshman

      Sophomore

      and so on

Keep your purpose and audience in mind when choosing a classification scheme. For instance, suppose you are classifying *insecticides.* For chemists, it might be most appropriate to classify insecticides by their chemical properties; however, for farmers, it might be best to classify according to the types of insects the insecticides control. You could classify *cities* in literally thousands of ways: by location; racial/ethnic mix; population; numbers of hotels and restaurants; type of government; availability, number, and size of convention rooms; and so forth. For convention planners, classification schemes based on the numbers of restaurants, hotels, and convention rooms might best serve their interests.

Mechanism Descriptions

Descriptions of mechanisms are common in technical writing. You will find many examples of them in technical advertisements, empirical research reports, and instructions. As always, the amount of detail presented should be based on your purpose, your readers' concerns and characteristics, your readers' level of knowledge and experience in the area, and your readers' attitudes toward your purpose. But in general, mechanism descriptions follow a three-part scheme. Here's an example, explaining a mechanism called a *scarifier,* which is used to prepare forest floors for the regeneration of trees:

1. Overview	The modified drag-chain is designed to be pulled by crawler-tractors in the 30- to 50-horsepower class. The modified drag-chain scarifier was designed to expose mineral soil in spot areas under standing trees. Preliminary tests indicate that the modified chain may distribute seed better than rakes or disks, although rakes and disks may provide better soil disturbance.
2. Division into component parts and description of the parts	The modified drag-chain employs two lengths of lightweight drag-chain instead of the three heavy strands in the original. Two-inch-square bar stock, 24 inches long, welded to each length of chain, increases scarification. . . .
3. Mechanism in action	The chain is self-cleaning and rolls over slash downfall better than other implements. Roots of competing grasses are pulled out by the chain. . . .[2]

Mechanism descriptions are generally accompanied by graphic illustrations, such as drawings and photographs (discussed further in Chapter 6).

Because similar situations occur so often in technical writing, many useful formats and organizational plans have been developed. You can often use such plans and formats, perhaps in modified forms, to choose and organize content in your writing. Part Two, The Formats of Technical Writing, will provide a lot of useful information on choosing and organizing content. Chapter 10, in particular, on formats of reports, will give you detailed accounts of how to write instructions, analytical reports, proposals, progress reports, and empirical research reports.

Knowing the basic formats and uses of these organizational schemes will help you organize a piece of writing. But nothing is as important as choosing your content and organizing it around your purpose and audience. In the HMO brochure example, knowledge of your purpose and audience should have led you to organize your writing by topic, even if you had never heard of topical arrangement.

Outlines

Make an outline when you are organizing. Writing things down helps clarify your thoughts. Things not written down may be forgotten. Trying to write something down and not being able to express it clearly may suggest to you that it's the wrong approach. You do not necessarily need to make a formal outline full of roman numerals and capital letters. But you should keep a good record of your organization with an informal outline of headings and subheadings. When you have a coherent outline that matches your purpose and audience, you will be ready to write.

ENDNOTES

[1]Adapted from Rebecca D. Williams, "FDA Proposes Folic Acid Fortification," *FDA Consumer* May 1994: 13.

[2]Adapted from Dick Karsky, "Scarifiers for Shelterwoods," *Tree Planters' Notes* 44 (1993): 14.

4

Write Clearly and Precisely

When you write the first draft of a document, do it rapidly and without much regard for mechanics and style. Using the content and organization you have arrived at (by following the guidelines in Chapters 1 through 3), get your information out where you can see it. Writing is thinking, so often, while writing, you will see different ways of organizing and different content choices. Follow the flow of your writing. Don't be a slave to organizational plans.

When you have completed your first draft, check its content and organization once again to be sure you have met your purpose and your audience's needs. Revise, if necessary. When you have done that, then it is time to make sure that your paragraphing, sentence structure, and language present your content clearly and precisely.

Paragraph for Readers

Imagine reading page after page of prose without any paragraphs. They would appear dense and forbidding. Therefore, the first principle of paragraphing is to *paragraph often,* so you don't frighten off your readers. Judging by standard practice in well-written prose, paragraphs of 60 to 100 words seem about right, depending in part on the page design.

For example, letters or pages that are formatted in narrow columns will likely have shorter paragraphs than standard report pages.

In technical writing, the first sentence in a paragraph generally introduces its subject and frequently provides a transition from the previous paragraph. You don't need to be heavy handed about either the introduction or the transition. These few paragraphs illustrate how to introduce the topic in the first paragraph and then move on to subtopics in subsequent paragraphs:

> The three major causes of land degradation are destructive agricultural practices, deforestation, and overgrazing.
>
> Destructive agricultural practices and land mismanagement account for 27 percent of the world's soil degradation, much of it in North America. Soil has been lost by repeated use of conventional tillage with heavy equipment and failure to use contour plowing on sloping terrain. Since 1930, the U.S. Government has spent $18 billion in conservation measures to reduce soil erosion. Despite present expenditures of $1 billion a year, the U.S. still annually loses some 6 billion tons of topsoil.
>
> Land degradation is especially acute in the former Soviet Union, a consequence of short-sighted industrial agricultural practices that ignored natural factors and faith that technology, fertilizers, and pesticides could increase crop yields interminably. Of Russia's 13.6 million acres of irrigated land, one-fifth is too salinized and two-fifths too acidified to support production. . . .
>
> Deforestation exposes fragile tropical soils to rainfall. . . .
>
> Overgrazing causes 35 percent of desertification, the most prevalent type of soil degradation. . . .[1]

The first one-sentence paragraph serves as a transition from the previous subject by announcing that the new subject will be *the three major causes of land degradation.* The first sentence in the second paragraph makes the transition to *destructive agricultural practices in North America.* The rest of the paragraph provides some supporting data.

The first sentence of the third paragraph lets you know that *land degradation* is still the subject but the focus has shifted to the former Soviet Union, and so on. The fourth paragraph, on *deforestation,* makes the transition to the new subject simply by beginning *Deforestation exposes . . .* The transition to *overgrazing* is made with the paragraph beginning *Overgrazing causes . . .*

The writer shifts gears between subjects simply by using key terms. The writer also keeps the reader's eyes on the subject by repeating key terms. Notice how many times *degradation* occurs in the sample paragraphs. (For that matter, notice how many times the words *paragraph* and *paragraphs* occur in this section discussing paragraphs.)

Not repeating key terms—or worse, using variant terms (for example, *deterioration* for *degradation*)—may cause your readers to lose sight of the subject or think a shift in subject has occurred. Don't be afraid to use intelligent repetition.

Use Language Appropriate for Your Readers

When you have done your audience analysis, you should have a good idea of the language level you can use. For instance, you will not want to throw terms from physics, sociology, or agronomy at readers who are not knowledgeable in those fields. On the other hand, for experts in these areas, more sophisticated language would be appropriate and even expected.

Here, for example, are a few sentences aimed at readers who are presumed to be knowledgeable about the words and concepts of *rhetoric* (which is the art of speaking or writing effectively). For the intended audience, the language used is entirely appropriate. But for other readers, at least four or five of the words and concepts presented will cause difficulty:

> A new model of the persuasive process follows from this understanding of persuasion as an essentially non-rational form of discourse directed at the listener's will. The new model is "motivistic" or "operational," monologic and mechanistic. The agonistic debate model is replaced by an image of an essentially manipulating persuader influencing a passive listener.[2]

Notice the difference in the language used in the next illustration, which is from a publication of the American Heart Association. This text is clearly aimed at intelligent but uninformed readers:

> When a heart attack occurs, the dying part of the heart may trigger electrical activity that causes *ventricular fibrillation*. This is an uncoordinated twitching of the ventricles that replaces the smooth, measured contractions that pump blood to the body's organs. Many times if trained medical professionals are immediately available, they can use electrical shock to start the heart beating again.
>
> If the heart can be kept beating and the heart muscle is not too damaged, small blood vessels may gradually reroute blood around blocked arteries. This is how the heart compensates; it's called *collateral circulation.*

For the most part, the American Heart Association sample uses simple, everyday words, such as *trigger, twitch,* and *blocked.* The two technical terms used—*ventricular fibrillation* and *collateral circulation*—are well defined.

Inflated language is never appropriate for any audience. Resist the temptation to impress your readers with fancy and pompous words. Choose simple, common words as much as possible. Don't *utilize* things; *use* them. Don't *initiate* and *terminate* things; *start* and *stop* them. Avoid phrases like *due to the fact that* and *at the present time;* simple words like *because* and *now* will serve you and your readers better.

Prefer the Active Voice

Active-voice sentences clearly state who or what the actor is and what the actor is doing. For that reason, most readers find sentences written in the active voice easier to follow and understand than those written in the passive voice. In addition, sentences written in the active voice seem more direct and interesting. You should use the active voice for the bulk of your writing.

In active-voice sentences, the subject acts in some way. For example:

The director reported that the spacecraft will begin mapping operations earlier than expected.

Phase I of the program runs until June.

The proposal includes two missions for 1999.

Notice that the subject does not have to be a person. As shown in these examples, the subject can be a *phase,* a *proposal*—anything at all.

In a passive-voice sentence, the subject is acted upon:

The ultraviolet emissions were detected by several astronomers.

Satellites, such as the earth's moon, are bound to their planets by the pull of gravity.

It is all too easy in a passive-voice sentence to omit the final prepositional phrase (beginning with *by*) that identifies the actor, even when that knowledge may be important. For example:

New technology was developed to revolutionize high-speed air travel.

To rewrite this sentence in the active voice requires adding an actor—who or what developed the new technology:

NASA developed new technology to revolutionize high-speed air travel.

Passive voice does have a place. When the identity of the actor is obvious or irrelevant (as is often the case in the "Materials and Methods" section of a research report), use the passive voice:

Seedling heights and ground line diameters were measured immediately after planting and at the end of one and two growing seasons. Percentage survival was calculated following the first and second growing season by dividing the number of surviving seedlings by the number of seedlings originally planted in each plot.[3]

Use the passive voice when it's appropriate, but prefer the active voice on most occasions. Your readers will appreciate it.

Use Personal Pronouns

Personal pronouns go hand in hand with the active voice. If the author or authors of a report wish to express an opinion or relate an action, it's appropriate to write an active-voice sentence that begins with *I* or *we,* as in:

> I recommend the opening of the new office in Dayton as soon as possible.

Not to use *I* in this sentence would result in a bloodless passive-voice sentence:

> The opening of a new office in Dayton is recommended by the author.

The following would be even worse, because now no one is accepting responsibility for the recommendation:

> The opening of a new office in Dayton is recommended.

Using the personal pronoun *you* in instructions clarifies the actor and personalizes the instructions, as in this passage from Internal Revenue Service (IRS) tax instructions:

> You can deduct the actual cost of running your car or truck or take the standard mileage rate. You must use actual costs if you do not own the vehicle or if you used more than one vehicle simultaneously (such as in fleet operations).

Take away the use of *you,* and the result would likely be the impersonal passive voice, harder to read and understand and vague about who is doing what:

> The actual cost of running a car or truck can be deducted or the standard mileage rate can be used. Actual costs must be used if the vehicle is not owned or if more than one vehicle is used simultaneously (such as in fleet operations).

In tax instructions, the IRS refers to itself as *we,* as in:

If you want, we will figure the tax for you.

The use of *we* in instructions is appropriate, as long as it is clear who *we* is. In the previous example, the use of *IRS* would have worked about as well as *we*. But *we* sounds much more friendly and personal, which may have a positive effect on the audience.

Often, you would be wise to make clear the *references* of your terms at the beginning of your instructions. For example, at the beginning of an insurance policy, you might note that *we* refers to the insurance company and *you* refers to the policy holder.

Use Action Verbs

Using action verbs is closely related to using active voice and personal pronouns. All these style decisions help you avoid using *nominalizations,* which are nouns derived from verbs. For example, the word *instruction* comes from *instruct,* and *assessment* comes from *assess.*

There is nothing wrong with using nominalizations, as long as they are used properly. Using the word *instructions* in a sentence like this is fine, for example:

Computer companies have learned that good instructions sell computers.

But you would be on soft ground and sinking fast if you wrote this passage in a letter to an office manager:

Misuse of the computer network by your secretaries has become evident. Please make provision for your secretaries to receive proper instruction in the use of the network.

This passage is stuffy, wordy, and not specific. This rewrite is better:

Your secretaries are misusing the computer network with personal mail. Please instruct them to use the network for business mail only.

In the revised passage, the writer uses action verbs and specifies what the problem actually is.

Unfortunately, you will find a good many nominalizations in technical writing, such as this one:

> The emission of sulfur dioxide from the factory is much greater than the emission of hydrogen sulfide.

If you find such a clumsy sentence in your work, think about where the action is and rewrite the sentence with an action verb:

> The factory emits much more sulfur dioxide than hydrogen sulfide.

When you are revising your work, look out for nominalizations. Better still, use the "find" or "search" function of your computer. Look for words that end in *-ment, -ion, -ance, -ence, -al,* and *-ing.* If you find any nominalizations, check to see if you have used them properly. If you have not, revise them by using action verbs.

Don't Introduce Unnecessary Complication

Both research and intuition will tell you that the more highly educated readers are, the more sentence complexity they can tolerate. But *no* readers appreciate sentences that are too long or tortured out of shape.

Think Subject-Verb

The subject and verb of a sentence are its frame, upon which you can hang various grammatical segments to expand or clarify the information provided by the subject and verb. To illustrate this process, here are a few examples:

> OPEC could easily produce half of all the oil consumed in the world.

> By 2010, OPEC could easily produce half of all the oil produced in the world.

> Because OPEC members control such a huge share of the world's high-quality, low-cost oil reserves, by 2010, OPEC could easily produce half of all the oil produced in the world.

None of these sentences should cause difficulty for a reader with at least high school reading ability. In all of them, the basic subject-verb frame stays clearly in view. The added grammatical segments provide additional information but do not excessively complicate matters.

Use a Reasonable Sentence Length

What is a *reasonable* sentence length? The answer to this question obviously relates closely to the abilities of your readers. Most high school graduates can read longer sentences that elementary schoolchildren, and most college-educated readers can read longer sentences than high school students. Moreover, readers familiar with the subject under discussion can read longer sentences than readers who are not.

As shown by the examples in the last section, sentences grow in length as information is added to them. At some point, they can be too long and too complicated, regardless of readers' abilities. The sentence in this example is probably too long for most readers:

> Because of OPEC members', especially the Persian Gulf members', control over such a huge share of the world's high-quality, low-cost oil reserves, the willingness and ability of OPEC members to expand production capacity, including production potential from Kuwait and Iraq, and the limited ability of non-OPEC producers to expand production facilities, by 2010, OPEC could easily produce half of all the oil produced in the world, which will greatly influence prospects for world oil prices.

A reader's ability to handle long, complicated sentences relates not only to the reader's own skill but also to the skill with which the writer constructs sentences. Therefore, it is risky to set limits on sentence length. It seems clear, however, that sentences that exceed 40 or 50 words are too difficult for most people. Professional writers average about 20 words a sentence. That average is likely one that most writers should strive for.

Write Positively

Too many negative words in a sentence can cause unnecessary complication, particularly in instructions. The problem occurs when negative words—such as *no, none,* and *not*—are combined with words that begin with negative prefixes—such as *ir-* (*irrelevant*), *non-* (*noncommittal*), and *un-* (*unbroken*). For example, the first sentence that follows is more difficult to read than the second:

> The virus protection is not installed properly until the virus protection icon appears at the bottom of your screen.

> The virus protection is installed properly when the virus protection icon appears at the bottom of your screen.

In reading the first sentence, readers will have a momentary pause while translating the negative combination *not . . . until* into a positive statement. The second sentence, already written positively, presents no such complication.

Negative combinations are particularly confusing in instructions, but they also make it difficult to understand the meanings of other types of sentences, as well. For example:

> The rebirth of the Cossack movement is not unpopular with the Ukrainian leadership.

Readers can interpret this sentence at least two ways: The leadership *likes* the rebirth, or the leadership *is neutral* about the rebirth. It's better not to waffle. Say what you mean in a positive way:

> The rebirth of the Cossack movement is popular with the Ukrainian leadership.

Better still:

> The Ukrainian leadership likes the rebirth of the Cossack movement.

Avoid Long Noun Strings

Using nouns to modify other nouns is commonplace in English. Expressions like *mail carrier* and *consumption level* are grammatically correct and understandable. But using long noun strings to modify other nouns introduces complications that raise difficulties for the reader, as in this example:

> Surplus production energy capacity price fluctuation control policies seem doomed to failure.

Policies is the word modified by the long noun string. That much is clear, but little else is. In the seven-word modifying phrase, the reader has to pause and sort out which words or groups of words modify other words or groups of words. The writer should do the sorting, perhaps in this way:

> The policies for controlling price fluctuations caused by surplus production in energy capacity seem doomed to failure.

Review your own writing to identify noun strings. Be wary if you see that you have put together more than three nouns to modify another noun without using clarifying hyphens or prepositions. You probably need to rewrite the sentence.

Check for Parallelism

Write *parallel* ideas in parallel grammatical forms. To do otherwise introduces complication, as in this example:

> Signs of a heart attack include a sensation of fullness, pain in the center of the chest, to faint, and when you feel a shortness of breath.

In this example, the writer started the list of symptoms with two noun phrases and then switched to an infinitive phrase, followed by a dependent clause. You really don't have to know all this grammatical terminology to recognize that the sentence has gone wrong somehow. It has become complicated and more difficult to understand than it should be.

Fix this sentence by putting all the symptoms in the same grammatical form:

> Signs of a heart attack include feeling a sensation of fullness, having pain in the center of the chest, fainting, and being short of breath.

Another situation in which to use parallelism is when joining two independent clauses to form a compound sentence. Specifically, do not make one clause active and the other passive, as in this example:

> Stroke victims tend to be slow and disorganized when faced with problems, and friends and family members are often surprised by their hesitant manner.

Even though this is an understandable sentence, it is awkward and needlessly complicated. Rewrite it this way, using parallel forms in both parts of the sentence:

> Stroke victims tend to be slow and disorganized when faced with problems, and their hesitant manner often surprises friends and family members.

Finally, keep grammatical forms parallel when making lists:

> Follow this advice to lower your risk of having a heart attack:
> - Have your blood pressure checked regularly.
> - Don't smoke.
> - Eat nutritious foods in moderate amounts.
> - Have regular medical checkups.

In this list, the writer used the imperative voice (i.e., "Do this . . .") for all four items. In addition, all the items are sentences. Finally, the use of bullets (•) helps clarify and organize the list items for readers.

ENDNOTES

[1]Nyle K. Walton, "Demographic Issues," *Geographic and Global Issues* Autumn 1993: 10–11.

[2]Arthur E. Walzer, "Rhetoric and Gender in Jane Austen's *Persuasion*," *College English* 57 (1995): 691.

[3]Adapted from J. L. Yeiser and E. J. Rhodenbaugh, "Early Survival and Growth of Loblolly Pine Seedlings Treated with Sulfometuron or Hexazinone Plus Sulfometuron in Southwest Arkansas," *Tree Planters' Notes* 45 (1994): 117.

5

*Use Good
Page Design*

Well-designed pages increase the accessibility of your report by helping readers see its organization. By making your presentation visually attractive, good design increases the likelihood that your audience will read carefully.

Important elements of good design are headings, headers and footers, appropriate type size and typeface, lists and informal tables, discreet typographical emphasis, and the ample use of white space.

Provide Headings

The use of headings lowers the density of type on the page and provides easy transitions from one topic to the next. Mainly, headings make information more accessible to readers.

Different levels of headings identify topics and subtopics within a document, making its organization clear to readers. By scanning headings or using them in conjunction with a matching table of contents (see pp. 86–87), readers can find the sections in documents they need or want to read. For example, an accountant reading a report about a new financial spreadsheet may be most interested in the how-to instructions. The accountant's boss, however, may be most interested in how the new spreadsheet may make the accountant more efficient.

Compare Figures 5.1 and 5.2 to see what a difference in readability a few well-placed headings can make.

Because the main role of headings is to increase accessibility, do not use more than three or four levels of headings. Too many headings chop a document into too many pieces and decrease, rather than increase, accessibility.

You will likely have two basic questions in using headings: (1) How do I phrase headings? and (2) How do I make headings noticeable and distinctive?

Phrasing Headings

Headings can be questions, short sentences, single words, or phrases of various types. For example, a document about government benefits might contain a section on qualifying for those benefits. The heading for this section could be phrased in various ways:

Who Can Qualify?

You May Qualify

Qualifications

Qualifying for Benefits

Questions seem to work well for headings, perhaps because they mirror what is in readers' minds. But the really important principle is that each heading should accurately identify what the section contains. For that reason, most headings should be substantive, rather than generic. A *generic* heading is a heading like "Part One," with no further identification. A generic heading such as "Introduction" or "Conclusion" serves the purpose adequately. But when used elsewhere in reports, generic headings do not give readers enough information. In the body of a report, use *substantive* headings that tell readers what they can expect to find in the sections, like these:

How English Began

Trends in Computer Use

Dangers in Self-Medication

The push for economic development and foreign investment in Vietnam, Laos, and Cambodia is threatening forests, waters, and wildlife. Deforestation is widely regarded as Indochina's greatest environmental problem.

About 70 percent of Laos was covered with trees in 1970, compared with less than 47 percent today. Vietnam's forests have shrunk from 44 percent in 1942 to about 24 percent today, and Cambodia's forests from 73 percent in 1960 to less than 40 percent. Laos, seen as Indochina's last "green" hope, still has large forest areas intact, but some logs—though officially banned from export—are among Laos's most valuable export commodities. In Vietnam, slash and burn agriculture in the highlands is out of control, even though it is illegal and doing untold damage to the environment.

Cambodia's Tonle Sap (Great Lake), one of the world's richest freshwater fishing grounds, suffers from sedimentation and loss of fish breeding grounds owing to forest and mangrove clearing on its banks. Fish catches, which reached an estimated 100,000 tons in 1970, have dropped in recent years to 70,000 tons or fewer per year. Laos views hydroelectric power as a valuable export, and officials are considering 25 power generating projects. Critics warn, however, that large dams would flood valuable forests, degrade now-pristine areas, and change water flows downstream.

In an attempt to balance economic with the cost of environmental development, Vietnam, Laos, and Cambodia are taking first steps to limit pollution and scarce resource exploitation. Despite the stated good intentions to address mounting environmental degradation, economic pressures to increase regional development will weaken governmental control and effectiveness, especially along border areas and in remote parts of Indochina. Successful implementation of environmental programs probably will take regional cooperation.

FIGURE 5.1   Document without Headings

Source: Adapted from Ray Lester, "Indochina: Environment at Risk," *Geographic and Global Issues* Spring 1994: 17–18.

Indochina: Environment at Risk

The push for economic development and foreign investment in Vietnam, Laos, and Cambodia is threatening forests, waters, and wildlife. Deforestation is widely regarded as Indochina's greatest environmental problem.

Dwindling Forests

About 70 percent of Laos was covered with trees in 1970, compared with less than 47 percent today. Vietnam's forests have shrunk from 44 percent in 1942 to about 24 percent today, and Cambodia's forests from 73 percent in 1960 to less than 40 percent. Laos, seen as Indochina's last "green" hope, still has large forest areas intact, but some logs—though officially banned from export—are among Laos's most valuable export commodities. In Vietnam, slash and burn agriculture in the highlands is out of control, even though it is illegal and doing untold damage to the environment.

Water Pollution

Cambodia's Tonle Sap (Great Lake), one of the world's richest freshwater fishing grounds, suffers from sedimentation and loss of fish breeding grounds owing to forest and mangrove clearing on its banks. Fish catches, which reached an estimated 100,000 tons in 1970, have dropped in recent years to 70,000 tons or fewer per year. Laos views hydroelectric power as a valuable export, and officials are considering 25 power generating projects. Critics warn, however, that large dams would flood valuable forests, degrade now-pristine areas, and change water flows downstream.

The Future

In an attempt to balance economic with the cost of environmental development, Vietnam, Laos, and Cambodia are taking first steps to limit pollution and scarce resource exploitation. Despite the stated good intentions to address mounting environmental degradation, economic pressures to increase regional development will weaken governmental control and effectiveness, especially along border areas and in remote parts of Indochina. Successful implementation of environmental programs probably will take regional cooperation.

FIGURE 5.2   Document with Headings

Source: Adapted from Ray Lester, "Indochina: Environment at Risk," *Geographic and Global Issues* Spring 1994: 17–18.

Headings within any section of a document must be grammatically parallel (see pp. 29–30). For example, all the major topic headings must be parallel. The subtopic headings within a major section must also be parallel, but they do not have to be parallel to subtopic headings in other major sections. The Contents for the two parts of this book makes this concept clear:

Part One	**The Seven Principles of Technical Writing**
1	Know Your Purpose
2	Know Your Audience
3	Choose and Organize Your Content around Your Purpose and Audience
4	Write Clearly and Precisely
5	Use Good Page Design
6	Think Visually
7	Write Ethically
Part Two	**The Formats of Technical Writing**
8	Purposes of Formats
9	Elements of Reports
10	Formats of Reports
11	Formats of Correspondence

The headings for the two parts are parallel noun phrases. (They are also substantive.) The chapter headings in Part One are all active/imperative sentences. The headings in Part Two are all noun phrases: parallel with each other but not with the headings in Part One.

Making Headings Noticeable and Distinctive

Good choices of words and grammatical forms help make headings noticeable and distinctive. Beyond that, good choices of typographical formats for your headings make them noticeable and distinguish different levels of headings.

WRITE CLEARLY AND PRECISELY

USE GOOD PAGE DESIGN

Write Clearly and Precisely

Use Good Page Design

Write clearly and precisely.

Use Good Page Design

Write Clearly and Precisely

Use Good Page Design

Write Clearly and Precisely

Use good page design.

FIGURE 5.3   Examples of Heading Styles

Figure 5.3 shows some of the various types of headings available to you. Some are available on typewriters and word processors and others, only on word processors. Notice that distinctiveness is obtained through such devices as underlining, spacing, italicizing, capitalizing, centering, indenting, using different type sizes, and so forth. However, all these headings are in the same typeface. Using different typefaces for different headings can lead to a gimmicky overkill.

Choose three or four of the available styles and stick with them through your document. Keep each heading with the section of text it identifies. Don't leave a heading dangling at the bottom of a page and start the section it identifies on the next page. Include at least two lines of the section with the heading. If you can't, move everything to the next page.

THE NEW STATES OF CENTRAL ASIA

KAZAKHSTAN
About four times the size of Texas . . . xxxxxxxxxxxxxxxxxxxx
xxxxxxxxxxxxxxxxxxxxxxxxxxxxxxxxxxxxxxxxxxxxxxxxxxxxxxxx
Hydrography
The surface water network . . . xxxxxxxxxxxxxxxxxxxxxxxxxxx
xxxxxxxxxxxxxxxxxxxxxxxxxxxxxxxxxxxxxxxxxxxxx
        Salt water. The saline Caspian Sea . . . xxxxxxxxxxxxxxxx
xxxxxxxxxxxxxxxxxxxxxxxxxxxxxxxxxxxxxxxxxxxxxxxx
        Fresh water. The rivers of . . . xxxxxxxxxxxxxxxxxxxxxx
xxxxxxxxxxxxxxxxxxxxxxxxxxxxxxxxxxxxxxxxx
Agriculture
Although only 15 percent of . . . xxxxxxxxxxxxxxxxxxxxxxxxxx
xxxxxxxxxxxxxxxxxxxxxxxxxxxxxxxxxxxxxxxxxxxxxxxxx

KYRGYZSTAN
A mountainous country slightly larger than Nebraska . . . xxx
xxxxxxxxxxxxxxxxxxxxxxxxxxxxxxxxxxxxxxxxxxxxxx

FIGURE 5.4   Examples of Headings Done on a Typewriter

Source: Adapted from Lee Schwartz and Leo Dillon, "Economic and Resource Issues," *Geographic and Global Issues* Summer 1993: 4.

Format all same-level headings in the same typographical style. See Figure 5.4 for a good selection using a typewriter only and Figure 5.5 for a good selection using a word processor.

Use Headers and Footers

Headers and footers are other ways of keeping readers on track. A *header* is a phrase that identifies the document or perhaps a section of it. As the name implies, a header appears at the head, or top, of the page.

THE NEW STATES OF CENTRAL ASIA

KAZAKHSTAN
About four times the size of Texas . . . xxxxxxxxxxxxxxxxxxxx
xxxxxxxxxxxxxxxxxxxxxxxxxxxxxxxxxxxxxxxxxxxxxxxxxxxxxxxx
Hydrography
The surface water network . . . xxxxxxxxxxxxxxxxxxxxxxxxxxx
xxxxxxxxxxxxxxxxxxxxxxxxxxxxxxxxxxxxxxxxxxxxxxx
   Salt water. The saline Caspian Sea . . . xxxxxxxxxxxxxxxx
xxxxxxxxxxxxxxxxxxxxxxxxxxxxxxxxxxxxxxxxxxxxxxxxxxxx
   Fresh water. The rivers of . . . xxxxxxxxxxxxxxxxxxxxxxx
xxxxxxxxxxxxxxxxxxxxxxxxxxxxxxxxxxxxxxxxx
Agriculture
Although only 15 percent of . . . xxxxxxxxxxxxxxxxxxxxxxxxx
xxxxxxxxxxxxxxxxxxxxxxxxxxxxxxxxxxxxxxxxxxxxxxxxxxxxxx

KYRGYZSTAN
A mountainous country slightly larger than Nebraska . . . xxx
xxxxxxxxxxxxxxxxxxxxxxxxxxxxxxxxxxxxxxxxxxxxxxxx

FIGURE 5.5   Examples of Headings Done on a Word Processor

Source: Adapted from Lee Schwartz and Leo Dillon, "Economic and Resource Issues," *Geographic and Global Issues* Summer 1993: 4.

A *footer* contains the same basic information but appears at the foot, or bottom, of the page.

In a short document, the title of the piece is usually presented in a header or footer. Sometimes, a header or footer may include the date and the name of the author of the document. In a longer document divided into chapters or major sections, the chapter or section heading is usually presented in a header or footer. Note that in this book, the part title is indicated on each left-hand page and the chapter title is indicated on each right-hand page.

Page numbers may appear with headers or footers or separately. Often, the information identifying the document or section will be in a header and the page number will be in a footer. Do not put an identifying header or footer on the title page or on the first pages of major sections, such as chapters. Do put page numbers on every page except the title page, however.

See Figure 5.6 for a selection of typical headers and footers.

Choose an Appropriate Type Size and Typeface

Deciding what is an *appropriate* type size and typeface most often means choosing type that eases the reader's task.

Type Size

Type sizes are expressed in units called *points.* The higher the number, or *point size,* the bigger the type:

9-point type

10-point type

12-point type

14-point type

18-point type

24-point type

In general, 10- to 12-point type is easy for most people to read. Use these sizes for the text in reports and correspondence. You may choose slightly larger type, such as 14-point, for headings to make them stand out from the text. Reserve type sizes from 18-point and up for brochures and other specialty items, for which you may want dramatic effects.

Geographic and Global Issues   98

xxxxxxxxxxxxxxxxxxxxxxxxxxxxxxxxxxxxxxxxxxxxxxxxxxxxxxx
xxxxxxxxxxxxxxxxxxxxxxxxxxxxxxxxxxxxxxxxxxxxxxxxxxxxxxx
xxxxxxxxxxxxxxxxxxxxxxxxxxxxxxxxxxxxxxxxxxxxxxxxxxxxxxx
xxxxxxxxxxxxxxxxxxxxxxxxxxxxxxxxxxxxxxxxxxxxxxxxxxxxxxx

Chapter 5   Demographic Issues

xxxxxxxxxxxxxxxxxxxxxxxxxxxxxxxxxxxxxxxxxxxxxxxxxxxxxxx
xxxxxxxxxxxxxxxxxxxxxxxxxxxxxxxxxxxxxxxxxxxxxxxxxxxxxxx
xxxxxxxxxxxxxxxxxxxxxxxxxxxxxxxxxxxxxxxxxxxxxxxxxxxxxxx
xxxxxxxxxxxxxxxxxxxxxxxxxxxxxxxxxxxxxxxxxxxxxxxxxxxxxxx

-99-

James Meadors
Progress Report
January 15, 1996

xxxxxxxxxxxxxxxxxxxxxxxxxxxxxxxxxxxxxxxxxxxxxxxxxxxxxxx
xxxxxxxxxxxxxxxxxxxxxxxxxxxxxxxxxxxxxxxxxxxxxxxxxxxxxxx
xxxxxxxxxxxxxxxxxxxxxxxxxxxxxxxxxxxxxxxxxxxxxxxxxxxxxxx
xxxxxxxxxxxxxxxxxxxxxxxxxxxxxxxxxxxxxxxxxxxxxxxxxxxxxxx

-4-

FIGURE 5.6   Typical Headers and Footers

Typeface

The two categories of type are serif and sans serif. *Serif* type has small extenders (called *serifs*) coming off the letters, as in the type used for most of the text in this book, which is Minion. Sans serif type does not have these extenders, as in Helvetica type, which is used for some of the tables and figures in this book.

Conventional design wisdom says that serif type is easier to read and that sans serif type is more modern looking. As done in this book, serif type is usually used for large sections of text and sans serif type is frequently used for graphs and tables.

Use Lists and Informal Tables

You can save words and open up your text by using lists and informal tables. Separate these elements from your text by indenting them at the left or both margins. Do not identify lists or informal tables with titles or table numbers.

A list with its accompanying text looks like this:

Some of the alternative therapies currently under study with grants from NIH include the following:

- acupuncture to treat depression, attention-deficit hyperactivity disorder, osteoarthritis, and post operative dental pain
- hypnosis for chronic low back pain and accelerated fracture healing
- biofeedback for diabetes, low back pain, and face and mouth pain caused by jaw disorders

While these alternative therapies are the subject of scientifically valid research, it's important to remember that at this time their safety and effectiveness are still unproven.[1]

Notice how bullets (•) are used to identify the list items. Numbers may also be used but only if the order of the items in the list is important.

In January and February of 1994, earthquakes caused 281 fatalities, as follows:

Iran	6
Indonesia (Sumetera)	207
Indonesia (Halmahera)	7
Uganda	4
USA (California)	57
Total	281

In Iran, three strong earthquakes struck in late February.

FIGURE 5.7   Informal Table

Source: Adapted from Waverly J. Person, "Earthquakes: January–February 1994," *Earthquakes & Volcanoes* 25 (1994): 52.

An informal table is essentially a list that contains columns (see Figure 5.7). Present informal tables as you do lists, but take care to align the separate columns of information. (See Chapter 6 for a discussion of formal tables.)

Use Emphasis Carefully

Use typographical variations to emphasize important points in your reports or correspondence. Variations include the following:

boldface
<u>underlining</u>
italic
larger type size
ALL CAPITAL LETTERS

When using any of these means of emphasis, be careful not to overdo it. A page with too many things emphasized will look cluttered, and the emphasis will be lost. Be particularly careful not to use all capital letters for more than one line. Because we rely on the ups and downs of capitals and lowercase letters in our reading, using all capitals raises the difficulty of reading. See Figure 5.8 for an example of typographical emphasis.

Leave Ample White Space

For ease of reading, a printed page should be about 50 percent type and 50 percent *white space* (a design term for "empty space"). Headings, lists, informal tables, and paragraphing all contribute to white space. In addition, you gain more white space with adequate margins, medium-length lines, and proper spacing. All measurements given here are based on using a standard 8-1/2″ x 11″ page.

Margins

Leave 1″ for the top and side margins and 1-1/2″ for the bottom margin. If you intend to bind your document, leave 1-1/2″ to 2″ on the side to be bound (usually the left).

Medium-Length Lines

Line length depends somewhat on the size and face of type used, but in most cases, a line of 50 to 70 characters (which is about 10 to 12 words) will be about right. In a double-column format, about 35 characters (or 5 words) per line is appropriate.

Should you *justify* your lines or not? That is, should your lines be even at both the left and right margins? Many word processors, in an attempt to justify the right margin, leave unattractive "rivers" of white space running through the text. Given this problem, you are better off in most instances to leave the right margin *ragged*—that is, unjustified. In any case, there seems to be little difference in readability between justified pages and ragged-right pages.

Proper Spacing

With most typewriters and word processors, you have the choices of single-space, space-and-a-half, and double-space. Lines in memos and letters are traditionally single-spaced, with double-spacing between

---

Global Enterprises

DATE   15 September 1996

TO   Roy Goss
              Ann Manchester
              Jim Morris
              Brittany Osborn
              Al Smith

FROM   Pat Macintosh

SUBJECT   Schedule for Reporting Monthly Meetings

Thank you all for agreeing to report our monthly meetings. What
follows is the schedule for the year:

1996	
October	Pat Macintosh
November	Jim Morris
December	Brittany Osborn
1997	
January	Roy Goss
February	Ann Manchester
March	Al Smith
April	Pat Macintosh
May	Jim Morris
June	Brittany Osborn
July	Roy Goss
August	Ann Manchester
September	Al Smith

**If you can't report the month for which you are scheduled, call
me and I'll arrange a switch. If I'm not available, call one of the
other reporters listed to take your place.**

Please use a memo format for your reports. Address them to
Dave Buehler, and send copies to Sally Barker and me.

FIGURE 5.8   Example of Typographical Emphasis within a Memo

paragraphs. To achieve more white space in other documents, use space-and-a-half or double-spacing. Double-space the first draft of any document to allow room for corrections, changes, and scribbling.

ENDNOTE

[1]Adapted from Isadora B. Stehlin, "An FDA Guide to Choosing Medical Treatments," *FDA Consumer* June 1995: 11.

6

Think Visually

As you think about and plan your documents, think visually as well as verbally. Graphics of various kinds play a major role in technical writing, often presenting data and ideas more efficiently and precisely than words. In technical writing, use graphics to show objects, processes, and data.

Showing Objects

The word *object* covers a lot of territory. It can mean machinery, mountains, tools, animals, ponds, glaciers, the inner ear—indeed, any material thing you can see or feel. Photographs and drawings are used to portray objects.

Photographs

Photographs have the advantage of realism. Using photographs to accompany a written mechanism description is an efficient means of ensuring that readers see the mechanism as it really is. Figure 6.1 shows how words and a photo work together for better understanding.

Photographs also can provide drama when it's needed or wanted. The photo in Figure 6.2 of a collapsed freeway span, brought down by

The modified drag-chain (see Figure 2) employs two lengths of lightweight drag-chain instead of the three heavy strands in the original. Two-inch-square bar stock, 24 inches long, welded to each length of chain increases scarification. Swivels divide the strands to provide a rolling action. Cross bars are 2 × 2 inches square × 20 inches long. During field tests, 10 links made up a strand that scarified a 6-foot-wide swath with about 50% scarification. Links weigh approximately 25 pounds each. The anchor chain scarifier weighs approximately 1,200 pounds. The scarifier incorporates a unique spread bar for use with a three-point hitch, which increases maneuverability. Hoist lines suspended from a pair of rigid arms raise and lower the unit.

Figure 2—*Modified anchor chain scarifier shown with three-point hitch drawbar and lift attachment.*

FIGURE 6.1   Words and Photograph Working Together

Source: Adapted from Dick Karsky, "Scarifiers for Shelterwoods," *Tree Planters' Notes,* 44 (1993): 14.

FIGURE 6.2   Collapsed Freeway Span

Source: James J. Mori, "Overview: The Northridge Earthquake: Damage to an Urban Environment," *Earthquakes and Volcanoes* 25 (1994): 7.

an earthquake, suggests the power of an earthquake in a way that would be difficult to achieve in words.

Photographs can be annotated, as in Figure 6.3. In general, annotations are written horizontally for ease of reading. Notice also the scale in Figure 6.3 used to indicate size. Such things as coins and actual rulers included in photographs also efficiently show scale.

Unless they are very carefully made, photographs have the disadvantage of including extraneous detail, as is obvious in Figure 6.1. The mushrooms in Figure 6.4 have been more carefully photographed, and extraneous detail has been held to a minimum.

FIGURE 6.3 Annotated Photograph. Dates and stars mark epicenters of three California earthquakes. A heavy white line marks surface rupture caused by a 1971 earthquake. The scale at the top of the photograph clarifies distances.

Source: James J. Mori, "Overview: The Northridge Earthquake: Damage to an Urban Environment," *Earthquakes and Volcanoes* 25 (1994): 13.

FIGURE 6.4   Photograph of Mushrooms. Careful photography will minimize background detail.

Source: Marian Segal, "Stalking the Wild Mushroom," *FDA Consumer* October 1994: 23.

Drawings

Drawings give you the advantage of control. You can eliminate extraneous detail and easily emphasize whatever you want to emphasize. You can do cutaways of objects that would be difficult or impossible to do in a photograph. Drawings also can be easily annotated. All these advantages are apparent in Figures 6.5 and 6.6.

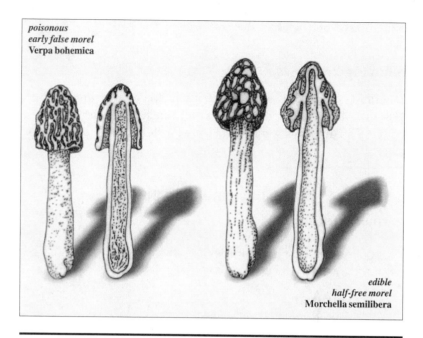

FIGURE 6.5   Cutaway Drawing of Mushrooms. The *left* annotations are on top and justified left and the *right* annotations are at the bottom and justified right, making for an excellent balanced design.

Source: Marian Segal, "Stalking the Wild Mushroom," *FDA Consumer* October 1994: 22.

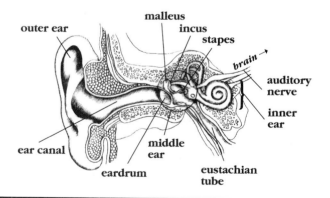

FIGURE 6.6   Cutaway Drawing of Ear. The usual horizontal orientation for the annotations is violated in an acceptable way to indicate the direction of the impulse to the brain.

Source: Rebecca D. Williams, "Protecting Little Pitchers' Ears," *FDA Consumer* December 1994: 12.

Showing Processes

Describing processes is a major activity in technical writing (see pp. 108–10). Many process descriptions will benefit from the descriptive power of accompanying graphics. Figure 6.7 shows how a combination

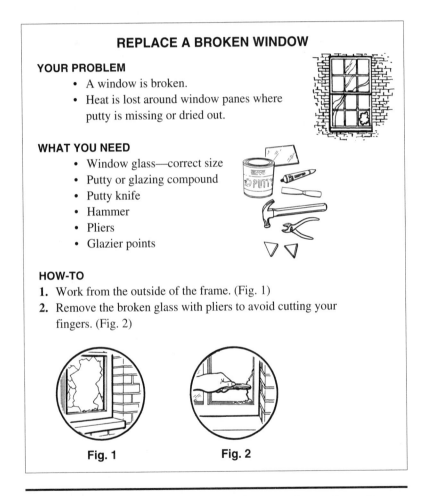

FIGURE 6.7 Words and Illustrations Working Together to Show a Process. The simple but excellent page design—using headings, bullets, and numbers—makes for easy reading.

Source: U.S. Department of Agriculture, *Simple Home Repairs: Inside* (Washington, DC: U.S. Government Printing Office, 1986) 11–12.

of words and illustrations work together to instruct readers how to accomplish a technical process.

Figures 6.8 and 6.9 illustrate *flowcharts:* graphs specifically designed to illustrate processes.

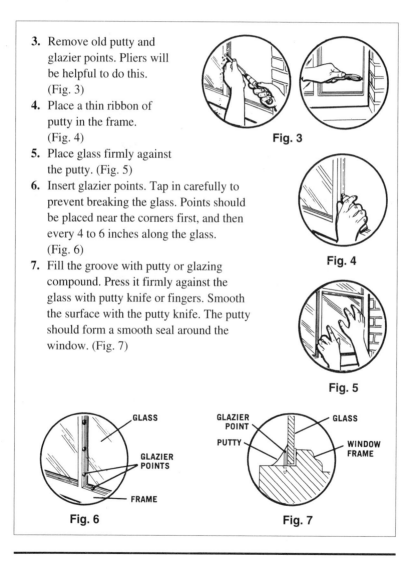

3. Remove old putty and glazier points. Pliers will be helpful to do this. (Fig. 3)
4. Place a thin ribbon of putty in the frame. (Fig. 4)
5. Place glass firmly against the putty. (Fig. 5)
6. Insert glazier points. Tap in carefully to prevent breaking the glass. Points should be placed near the corners first, and then every 4 to 6 inches along the glass. (Fig. 6)
7. Fill the groove with putty or glazing compound. Press it firmly against the glass with putty knife or fingers. Smooth the surface with the putty knife. The putty should form a smooth seal around the window. (Fig. 7)

Fig. 3

Fig. 4

Fig. 5

GLASS

GLAZIER POINTS

FRAME

Fig. 6

GLAZIER POINT

PUTTY

GLASS

WINDOW FRAME

Fig. 7

FIGURE 6.7   Continued

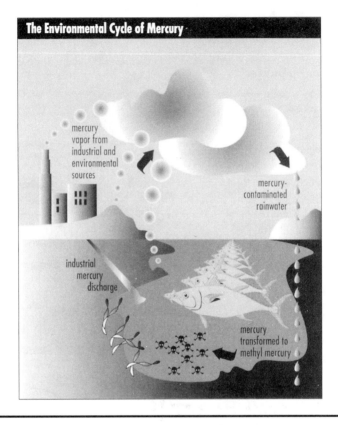

FIGURE 6.8   Flowchart. Flowcharts are suitable for all levels of readers but are particularly good for showing processes and cycles to nontechnical readers.

Source: Judith E. Foulke, "Mercury in Fish: Cause for Concern?" *FDA Consumer* September 1994: 7.

Showing Data

Tables and graphs of all sorts are great word and space savers in displaying data. With tables and graphs, you can summarize data and show trends and relationships among them.

Tables

Informal tables work well for simple data displays (see pp. 42–43). But for larger and more complex displays, you'll need formal tables. The sample table in Figure 6.10 illustrates a formal table. The accompanying

Processing of Milk Treated with rbST

Milk from rbST-treated and untreated cows is collected in the same manner. Milk from each farm is tested for antibiotic drug residues. If there are unsafe drug residues, the entire tanker of milk is dumped. If no residues are found the tanker delivers the milk to the processor who readies it for market. Antibiotics are used to treat mastitis, an inflammation of the cow's udder, which is more common in rbST-treated cows.

FIGURE 6.9   Flowchart. Flowcharts can be used to show decision points in a process, as demonstrated by this chart.

Source: Kevin L. Ropp, "New Animal Drug Increases Milk Production," *FDA Consumer* May 1994: 26.

Table number and title → **No. 663. Average Annual Pay, by State: 1991 and 1992**

Unit indicator / *Explanatory headnote* / *Footnote indicator*

[In dollars, except **percent change**. For workers covered by State unemployment insurance laws and for Federal civilian workers covered by unemployment compensation for Federal employees, approximately 90 percent of total civilian employment. Excludes most agricultural workers on small farms, all Armed Forces, elected officials in most States, railroad employees, most domestic workers, employees of certain nonprofit organizations and most self-employed individuals. Pay includes bonuses, cash value of meals and lodging, and tips and other gratuities]

Column heads and subheads / *STATE* / *Stub heading*

STATE	AVERAGE ANNUAL PAY 1991	AVERAGE ANNUAL PAY 1992[1]	Percent change, 1991-92[1]	STATE	AVERAGE ANNUAL PAY 1991	AVERAGE ANNUAL PAY 1992[1]	Percent change, 1991-92[1]
United States	24,578	25,903	5.4	Missouri	22,574	23,550	4.3
Alabama	21,287	22,340	4.9	Montana	18,648	19,378	3.9
Alaska	30,830	31,825	3.2	Nebraska	19,372	20,355	5.1
Arizona	22,207	23,161	4.3	Nevada	23,083	24,743	7.2
Arkansas	19,008	20,108	5.8	New Hampshire	23,600	24,925	5.6
California	27,513	28,934	5.2	New Jersey	29,991	32,125	7.1
Colorado	23,981	25,040	4.4	New Mexico	20,272	21,051	3.8
Connecticut	30,689	32,587	6.2	New York	30,011	32,399	8.0
Delaware	25,647	26,596	3.7	North Carolina	21,095	22,248	5.5
District of Columbia	35,570	37,971	6.8	North Dakota	18,132	18,945	4.5
Florida	21,992	23,144	5.2	Ohio	23,602	24,846	5.3
Georgia	23,165	24,373	5.2	Oklahoma	20,968	21,699	3.5
Hawaii	24,104	25,613	6.3	Oregon	22,338	23,514	5.3
Idaho	19,688	20,649	4.9	Pennsylvania	24,393	25,785	5.7
Illinois	26,317	27,910	6.1	Rhode Island	23,082	24,315	5.3
Indiana	22,522	23,570	4.7	South Carolina	20,439	21,423	4.8
Iowa	19,810	20,937	5.7	South Dakota	17,143	18,016	5.1
Kansas	21,002	21,982	4.7	Tennessee	21,541	22,807	5.9
Kentucky	20,730	21,858	5.4	Texas	23,760	25,080	5.6
Louisiana	21,503	22,340	3.9	Utah	20,874	21,976	5.3
Maine	20,870	21,808	4.5	Vermont	21,355	22,347	4.6
Maryland	25,962	27,145	4.6	Virginia	23,805	24,937	4.8
Massachusetts	28,041	29,664	5.8	Washington	23,942	25,553	6.7
Michigan	26,125	27,463	5.7	West Virginia	21,356	22,169	3.8
Minnesota	23,962	25,315	5.6	Wisconsin	21,838	23,022	5.4
Mississippi	18,411	19,237	4.5	Wyoming	20,591	21,215	3.0

Notes →

[1]Preliminary

Source: U.S. Bureau of Labor Statistics, *Employment and Wages Annual Averages 1992*; and USDL News Release 93-371, *Average Annual Pay by State and Industry, 1992.*

FIGURE 6.10 **Annotated Formal Table.** The annotations label the key parts of a table.

56

annotations describe the table's main features. In all the sample tables, notice that whole numbers are lined up on the last digits and fractional numbers, on the decimals.

The table in Figure 6.10 is complex enough that its lines serve the useful purpose of separating the data for easy reading. In less complex tables, many of the lines are eliminated (as in Figures 6.11 and 6.12), and white space serves to separate the data. Let ease of reading the table be your guide, but eliminate as much clutter as possible.

When need be, you can interpret your data in headnotes, captions, and annotations. Many of the figures in this chapter illustrate how this is done. Often, you will interpret data in the text accompanying your figures. Figure 6.12 illustrates this table-text relationship.

Table 2—*Survival of outplanted loblolly pines removed from* Macrophomina phaseolma-*infested seedbeds after 1 and 2 growing seasons in the field*

	% Survival			
	Sept. 2, 1988		Sept. 21, 1989	
Rep. no.	Trt I*	Trt II*	Trt I*	Trt II*
1	97.9	89.8	89.6	87.8
2	98.0	93.9	94.0	85.7
3	100.0	88.5	91.7	82.7
4	100.0	92.2	90.0	88.0
5	100.0	97.9	94.0	89.6
Mean‡	99.2a	92.5b	91.9a	86.8b

*Trt I = healthy seedlings from unaffected portions of seedbeds; Trt II = "healthy" seedlings removed from portions of seedbeds with abundant seedling mortality.

‡Means across computed sampling dates that are followed by different letters differ significantly at $P \leq 0.05$.

FIGURE 6.11 Simple Table. Use no more lines in a table than are necessary to make data easily readable.

Source: E. L. Barnard, "Nursery-to-Field Carryover and Post-Outplanting Impact of *Macrophomina phaseolina* on Loblolly Pine on a Cutover Forest Site in North Central Florida," *Tree Planters' Notes* 45 (1994): 70.

Cuba's total labor force, estimated at 4.3 million in 1990, will grow to only 5.5 million by 2010, assuming there is no migration from the island. This 26-percent increase is far below the 67-percent growth projected for the region as a whole. It would be further reduced by emigration, the impact of which can be estimated by making assumptions about its size. If a constant total net emigration of 25,000 per year is assumed, the increase would be only 19 percent, to about 5.2 million (see Table 1, below).

Table 1: Projected Cuban Labor Force, All Ages (millions)

	No Net Emigration			Constant Net Emigration 25,000/Year		
	Male	Female	Total	Male	Female	Total
1990	2.9	1.4	4.3	2.9	1.4	4.3
1995	3.2	1.6	4.8	3.2	1.6	4.8
2000	3.4	1.8	5.2	3.3	1.7	5.0
2005	3.5	1.8	5.3	3.3	1.8	5.1
2010	3.6	1.9	5.5	3.3	1.8	5.2

The Age Divide

The more striking story appears when the labor force is divided by age into two groups: 15–39 and 40–64. The younger portion of the labor force will increase by less than 4 percent by 2010 under the no emigration assumption, and it would actually *decrease* by 3 percent under the constant migration scenario described above (see Table 2, below). The decline is even greater among males. The proportionate share of the younger labor force will slip from 66 percent to 54 percent, and the median age of the whole labor force—already the oldest in the region—will rise from 33.5 in 1990 to 38.9 by 2010.

Table 2: Projected Cuban Labor Force, Age 15–39 (millions)
(assuming constant net emigration 25,000/year)

	Male	Female	Total*
1990	1.8	1.0	2.9
1995	2.0	1.1	3.1
2000	2.0	1.2	3.2
2005	1.9	1.2	3.0
2010	1.7	1.1	2.8

*May not equal sum of male and female components owing to rounding.

FIGURE 6.12   Three Tables with Accompanying Text. Tables display data well, but often, they must be explained and interpreted in accompanying text.

Source: Adapted from David G. Smith, "Cuba's Approaching Youth 'Bust,'" *Geographic and Global Issues* Spring 1994: 9–10.

Graphs

Bar, pie, line, and map graphs are all commonly used in technical writing. Pictographs are sometimes used but mainly for nontechnical audiences. All graphs can be used to summarize and to show trends and relationships. Bar and pie graphs show well the relationships among data. They are good for all levels of audiences, technical and nontechnical.

Line graphs are superior to bar graphs in showing the shapes of data. Are the numbers increasing? decreasing? forming a bell curve? Line graphs show these trends well, but be mindful of your audience. Line graphs work well for technical audiences, but unless you keep the line graph simple, nontechnical audiences may have trouble reading them. Consider using map graphs when there is a geographical component to your data. As you do with tables, provide whatever interpretation is needed on the graph or in the text.

Figures 6.13 through 6.18 illustrate various kinds of graphs in use. The figure captions point out various graph features and principles of their use.

Principles of Tables and Graphs

Graphs and tables should be:

- Clear, uncluttered, and efficient
- Suited to their readers
- Interpreted as needed with notes, captions, annotations, lines, keys, arrows and text (Footnotes are internal to tables and graphs and marked by numbers, letters, or symbols, such as asterisks [*].)
- Placed near their textual references
- Referred to when needed
- Numbered and given succinct titles: "Average Annual Pay, by State: 1991 and 1992" not "A Summary of Average Annual Pay, by State: 1991 and 1992"
- Well made and esthetically pleasing but not artsy (Too much decoration gets in the way of the message.)
- Legible
- Honest and truthful (see pp. 69–71)

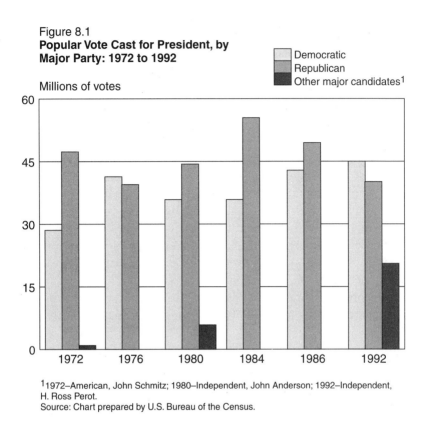

Figure 8.1
Popular Vote Cast for President, by
Major Party: 1972 to 1992

☐ Democratic
☐ Republican
■ Other major candidates[1]

Millions of votes

[1]1972–American, John Schmitz; 1980–Independent, John Anderson; 1992–Independent, H. Ross Perot.
Source: Chart prepared by U.S. Bureau of the Census.

FIGURE 6.13   Bar Graph. When color isn't an option, use different degrees of shading to distinguish among bars. Crosshatching, as shown in Figure 6.14, is not a good choice because the moiré effect it produces can distract readers. Avoid clutter of any kind. Use a complete grid behind the bars only when the reader will need to pick precise data off the graph. Notice that because the highest bar is in the 50 million range, the graph only goes to 60 million. Do not waste space in graphs.

Source: U.S. Bureau of the Census, *Statistical Abstract of the United States,* 114th ed. (Washington, DC: U.S. Government Printing Office, 1995) 268.

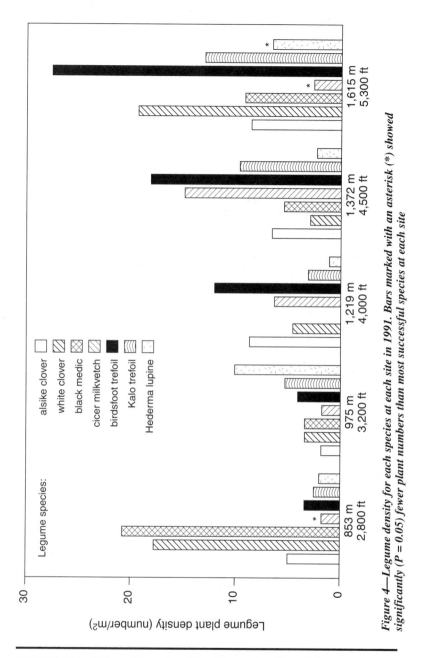

Figure 4—Legume density for each species at each site in 1991. Bars marked with an asterisk (*) showed significantly (P = 0.05) fewer plant numbers than most successful species at each site

FIGURE 6.14 Effect of Crosshatching. The moiré effect of the crosshatching decreases the readability of this graph. Instead of crosshatching, use different degress of shading, as in Figure 6.13.

Source: B. Java, R. Everett, T. O'Dell, and S. Lambert, "Legume Seedling Trials in a Forested Area of North-Central Washington," *Tree Planters' Notes* 46 (1995): 23.

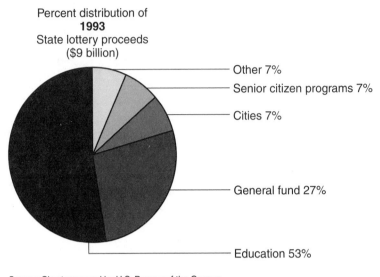

Source: Chart prepared by U.S. Bureau of the Census.

FIGURE 6.15   Pie Graph. In constructing a pie graph, use a rational order for your slices, generally large to small or small to large. Keep your labels horizontal to the page. When there is room, you may label inside the slice, but keep the lettering horizontal.

Source: U.S. Bureau of the Census, *Statistical Abstract of the United States*, 114th ed. (Washington, DC: U.S. Government Printing Office, 1995) 296.

FIGURE 6.16   Two Typical Line Graphs. The plotted lines in each graph are ➤ labeled directly on the graph, using lettering horizontal to the page, rather than using a key. Individual lines are kept distinct by using a mixture of dots and dashes and different thicknesses of solid lines. The fertility rate graph (*top*) uses no grid at all, and the consumer complaints graph (*bottom*) uses only a light grid. In both cases, the trend of the data is more important than precise reading.

Sources: (*Top*) Michael S. Owen, "Pakistan: Ramifications of Population Growth," *Geographic and Global Issues* Spring 1993: 21; and (*Bottom*) U.S. Bureau of the Census, *Statistical Abstract of the United States,* 114th ed. (Washington, DC: U.S. Government Printing Office, 1995) 646.

**Total Fertility Rates for
Selected Countries, 1970–92**
(Average number of children born per female)

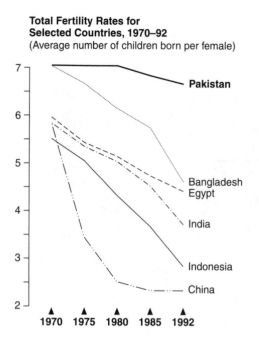

**Consumer Complaints Against
U.S. Airlines: 1986 to 1993**

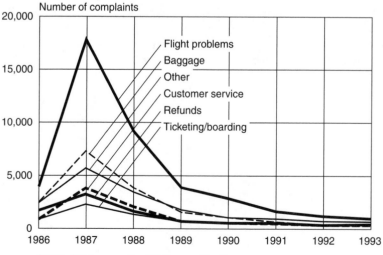

Source: Chart prepared by U.S. Bureau of the Census.

Figure 1.1
Projected Percent Change in State Populations: 1990 to 2000

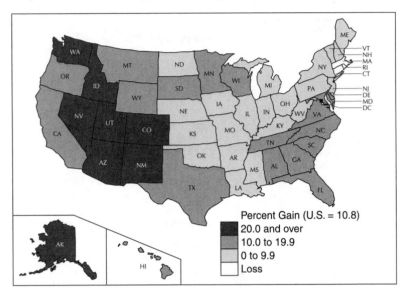

Source: Chart prepared by U.S. Bureau of the Census. Based on preferred series A.

FIGURE 6.17   Map Graph. Map graphs are very effective when geography is a factor, as it is in this graph that projects trends in state populations. Different degress of shading, not crosshatching, distinguish among the different percentage changes.

Source: U.S. Bureau of the Census, *Statistical Abstract of the United States,* 114th ed. (Washington, DC: U.S. Government Printing Office, 1995) 7.

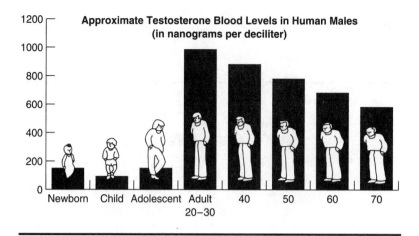

FIGURE 6.18   Pictograph. Pictographs are essentially bar graphs with human interest; they are used primarily for nontechnical audiences. Be sure the things depicted—in this case, children and men—can be easily recognized.

Source: Ken Flieger, "Testosterone: Key to Masculinity and More," *FDA Consumer* May 1995: 31.

Pre-proposal

7

Write Ethically

Ethical systems, whether religious or philosophical, agree that it is un-ethical to lie, cheat, and steal. Some systems grant a few exceptions, such as white social lies, a mother stealing for her starving child, and lying to the enemy in time of war. But in general, the agreement is universal.

Morality consists of being ethical when it would be safer, more convenient, and more profitable to be unethical. Thus, morality does not always come easily. People are often tempted to commit unethical acts for personal gain, out of loyalty to organizations, or out of fear of the consequences they will face if seen as being disloyal—a "whistle blower," for example.

Technical writing has consequences. On the basis of feasibility studies and proposals, governments and businesses spend millions, even billions, of dollars. People follow instructions, expecting that their safety or the safety of their equipment will not be compromised by misstatements. Scientists base future research on past research reports. Researchers who misrepresent results or tell outright lies in their reports can mislead other scientists for years. Therefore, a moral imperative exists for technical and scientific writers to write ethically.

The principles that follow tell you how to write ethically. They cannot make you act morally. Only you can do that.[1]

66

Don't Hide or Suppress Unfavorable Data

Imagine that you are conducting a feasibility study (see pp. 113–18) that will determine if your company will open a branch in Tampa, Florida. You have reason to believe that if the branch proves feasible, you will manage it. You really want to manage the branch and live in Tampa. Suppose, however, that some of the data you gather in your study do not support the new branch. Perhaps you find that transportation to the primary markets for your company's product will be too expensive from the Tampa location. Your temptation will be either to suppress the data altogether or hide them in some little-read location in the report, perhaps an appendix. Either action would be tantamount to lying and would be unethical behavior.

Similar situations may arise in writing proposals and research reports (see pp. 119–22 and 124–25). In a proposal, the temptation is to hide material that would indicate your company is not suited for the work it proposes to do. In a research report, the data might show that your theory is not as sound as you think. In both situations, the temptation is to hide or suppress the data.

Obviously, the people who read your reports, whatever kinds of reports they may be, put their implicit trust in your writing honest, complete reports. To violate that trust would be to act immorally.

Don't Exaggerate Favorable Data

Exaggerating favorable data is the reverse of suppressing unfavorable data. In writing a proposal, you might exaggerate the experience of your company's scientists, making them sound more expert than they really are. In a feasibility report or the analysis section of a research report, to support the decision or conclusion you want, you might give favorable data more weight than they deserve.

Is any such exaggeration ever ethical? Where proposals and other sales documents are concerned, the expression "Put your best foot forward" applies. That is, it is legitimate in advertisements and proposals to show how your product or service meets the needs of potential customers. You may do so by emphasizing the strong points of your product or service. You are *expected* to do so by both your organization and your customer. But to be ethical, such emphasis must not overstep the bounds and distort the true facts. For example, in a proposal, you could legitimately emphasize the Ph.D. in biochemistry held by your lead investigator, but it would be unethical to imply that her experience matched the needs of the client if, in fact, it did not.

In nonsales documents, such as research reports and feasibility reports, anything less than the relevant data, accompanied by an objective analysis, would be unethical.

Don't Make False Implications

In making a *false implication,* you are actually telling the truth but in a way that leads readers to the wrong conclusion. For example, imagine that you are writing a proposal for construction work in which safety on the job is of major importance. For eight years, your company had an enviable safety record, with an accident rate far below the industry average. However, in the last two years, because the company has not upgraded the equipment used by your employees, the accident rate has soared above the industry average. Even so, the average rate for the ten years is still slightly below the industry average.

Given this, you could truthfully make the statement "Our average accident rate over the last ten years has been below the industry average." But in doing so, you would be falsely implying that your present operations are being conducted safely. You would be making an unethical statement. Were you to say "Our average accident rate over the last ten years has been *substantially* below the industry average," you would be adding the sin of exaggerating favorable data. Benjamin Franklin had it right when he said, in *Poor Richard's Almanack,* "Half the truth is often a great lie."

Don't Plagiarize

To *plagiarize* is to take the words or ideas of others and present them as your own. Much technical writing is based on research into other people's writing. It is legitimate to use other people's data and ideas, but you must give appropriate credit (see pp. 100–105). It is not legitimate to present the words and sentences of others as your own. You must quote, paraphrase, or summarize.

Seeming exceptions to this principle sometimes occur in technical writing. For example, organizations that write many proposals have large blocks of material available for use by proposal writers, such as descriptions of company facilities and equipment. Because this material, often known as "boilerplate," belongs to the organization and not the original writers, it can be used legitimately without attribution.

Construct Ethical Graphs

Like words, graphs can lie, suppress, exaggerate, and tell half-truths. The basic rule for integrity in graphs is that the physical representation of the data must accurately reflect them.[2] For example, if the number of accidents in a plant has increased and decreased only slightly over the years, the curve on the graph representing those changes should be very shallow. However, by drawing a narrow graph, a graphic artist can end up with a steep curve and thus misrepresent the changes. See Figure 7.1 for an example of an unethical bar graph.

Numbers that change in only one direction can be misrepresented by changing the physical dimensions of the graph in two directions. For example, in a bar graph, increasing the sizes of the bars both vertically and horizontally will increase the area of the bars out of proportion to the actual increases in data, thus greatly exaggerating them. Pictographs that portray physical objects, such as people and factories, often lack integrity because they increase in two or three dimensions while the underlying numbers increase in only one (see Figure 7.2).

Because of the devaluation of the dollar caused by inflation, graphing in *current* dollars can distort the true growth in prices, wages, and such figures as the federal debt. In dealing with dollars, use *constant* dollars. In constant dollars, the value given for the dollar for a specific year is 1. All dollar values for years before and after the chosen year are then valued in proportion to the constant in a way that reflects inflation. For example, a dollar value for a year before the chosen year may have a value of 1.041, and one after may have a value of 0.984. The table in Figure 7.3 shows how this sysytem works. The graph in Figure 7.4

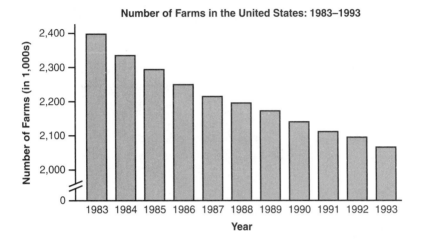

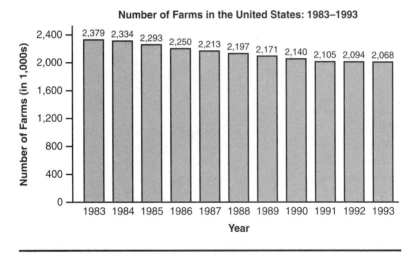

FIGURE 7.1   Vertical Misrepresentation of the Data. The *top* graph greatly exaggerates the decline in the number of farms in the United States for the years 1983 to 1993. The *bottom* graph represents the data accurately. Also in the bottom graph, labeling each bar with the appropriate number increases the integrity of the graph.

Source: Data from U.S. Bureau of the Census, *Statistical Abstract of the United States,* 114th ed. (Washington, DC: Government Printing Office, 1994) Table 1085.

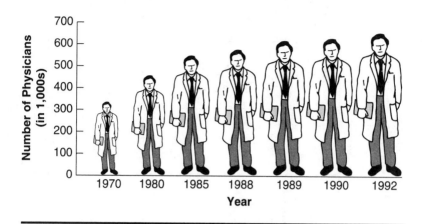

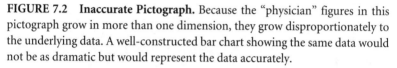

FIGURE 7.2 Inaccurate Pictograph. Because the "physician" figures in this pictograph grow in more than one dimension, they grow disproportionately to the underlying data. A well-constructed bar chart showing the same data would not be as dramatic but would represent the data accurately.

Source: Data from U.S. Bureau of the Census, *Statistical Abstract of the United States,* 114th ed. (Washington, DC: Government Printing Office, 1994) Table 171.

demonstrates the inaccuracy of graphing in current dollars compared to graphing in constant dollars.

Inexperienced graph readers can be easily misled by unethical graphs. Experienced graph readers will spot graphs that misrepresent data and, therefore, mistrust the author of the report. You owe it to yourself and to your readers to graph ethically.

Don't Lie

Many of the previous principles deal with unethically shading the truth. The final principle is all encompassing: *Do not lie.* Scientists and technicians are overwhelmingly honest, but there are exceptions. In this century, a scientist has fabricated case histories to support his psychological theories. A few scientists have plagiarized the works of others. These few

[Indexes: PPI, 1982=$1.00; CPI, 1982–84=$1.00. Producer prices prior to 1961, and consumer prices prior to 1964, exclude Alaska and Hawaii. Producer prices based on finished goods index. Obtained by dividing the average price index for the 1982=100, PPI; 1982–84=100, CPI base periods (100.0) by the price index for a given period and expressing the result in dollars and cents. Annual figures are based on average monthly data]

Year	Annual Average as Measured by—		Year	Annual Average as Measured by—		Year	Annual Average as Measured by—	
	Producer prices	Consumer prices		Producer prices	Consumer prices		Producer prices	Consumer prices
1950	$3.546	$4.151	1965	2.933	3.166	1980	1.136	1.215
1951	3.247	3.846	1966	2.841	3.080	1981	1.041	1.098
1952	3.268	3.765	1967	2.809	2.993	1982	1.000	1.035
1953	3.300	3.735	1968	2.732	2.873	1983	0.984	1.003
1954	3.289	3.717	1969	2.632	2.726	1984	0.964	0.961
1955	3.279	3.732	1970	2.545	2.574	1985	0.955	0.928
1956	3.195	3.678	1971	2.469	2.466	1986	0.969	0.913
1957	3.077	3.549	1972	2.392	2.391	1987	0.949	0.880
1958	3.012	3.457	1973	2.193	2.251	1988	0.926	0.846
1959	3.021	3.427	1974	1.901	2.029	1989	0.880	0.807
1960	2.994	3.373	1975	1.718	1.859	1990	0.839	0.766
1961	2.994	3.340	1976	1.645	1.757	1991	0.822	0.734
1962	2.985	3.304	1977	1.546	1.649	1992	0.812	0.713
1963	2.994	3.265	1978	1.433	1.532	1993	0.802	0.692
1964	2.985	3.220	1979	1.289	1.380			

FIGURE 7.3   Purchasing Power of the Dollar: 1950–1993

Source: U.S. Bureau of the Census, Statistical Abstract of the United States, 114th ed. (Washington, DC: Government Printing Office, 1994) Table 746.

Note: Data from U.S. Bureau of Labor Statistics. Monthly data in U.S. Bureau of Economic Analysis, Survey of Current Business.

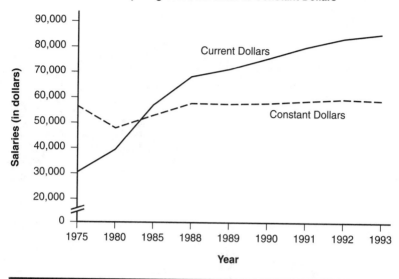

FIGURE 7.4 Comparison of Current Dollars to Constant Dollars. The curve using "Current Dollars" gives the false impression that high school superintendents have enjoyed almost a threefold increase in buying power over an 18-year period. The curve using "Constant Dollars" accurately shows that the increase has been quite modest.

Source: Data from U.S. Bureau of the Census, *Statistical Abstract of the United States,* 114th ed. (Washington, DC: Government Printing Office, 1994) Table 247.

have been outright liars. Science and technology are built on trust. To violate that trust is to shake the very foundation on which science and technology are built.

ENDNOTES

[1]The ethical principles in this chapter are based on those expressed in R. John Brockman and Fern Rook, eds., *Technical Communication and Ethics* (Washington, DC: Society for Technical Communication, 1989), in particular, H. Lee Shimberg's "Ethics and Rhetoric in Technical Writing," 59–62.

[2]For an excellent discussion of the principles given in this section, see Edward R. Tufte, *The Visual Display of Quantitative Information* (Cheshire, CT: Graphics Press, 1983) 53–87.

PART TWO

The Formats
of Technical Writing

8

Purposes of Formats

Format, in technical writing, refers to two things:

- The elements of a report, such as tables of contents, summaries, prefaces, introductions, and glossaries.

- The organization and sections of certain kinds of reports—For example, the format of an empirical research report written for fellow scientists will generally include the following sections, in this order: abstract, introduction, literature review, materials and methods, results, discussion, and conclusions.

Chapter 9 discusses the elements of reports. Chapter 10 discusses the formats of reports, covering the organizations and sections of several major kinds of reports. Chapter 11 discusses the formats of correspondence.

Despite seeming arbitrary at times, formats are very functional. For the reader, they improve accessibility, selective reading, and comprehension. For the writer, formats aid in developing the subject logically. Further, the writer's knowing what each element comprises makes it more likely that each element will be a useful part of the report.

A discussion using the empirical research report as an example illustrates each of these points. In following the discussion, remember that the readers of empirical research reports are generally already familiar with the vocabulary and concepts of the field.

Accessibility

Custom requires that each section of a research report be clearly labeled with a heading: *Abstract, Introduction,* and so forth. Through regular use of such reports, readers grow familiar with the content and purpose of each section and can quickly access whatever section of the report they wish to read.

Selective Reading

With accessibility comes the ability to read selectively. A well-written *abstract* summarizes the key points of the report. After reading it, the reader will know the objective of the research, the major results, the meaning of the results, and the major conclusions of the author. The reader now has many options: Read the entire report, read only the discussion to see how the author analyzes the results, check out the conclusions, and so forth. The reader's knowledge, needs, and interests—not the author's—govern the options chosen.

Comprehension

To understand and to gain new knowledge, the reader has to be led from the known to the unknown—that which is to be learned. Most formats allow for this logical progression. In the case of the empirical research report, the *introduction* and *literature review* together take the reader from the known to the unknown. By reviewing past research in the field, the literature review provides the necessary background for the reader to understand the objective of the research, the need for the research, and the techniques used in the research.

With this new knowledge, the reader can follow the descriptions of the materials and methods used in the research. And with that understanding, the reader can comprehend the results achieved. With all the preceding known, the reader can follow the analysis of the results and understand the conclusions. In sum, the author has led the reader from the known to the unknown and made the unknown known.

The progression from the known to the unknown also supports selective reading. For instance, the reader may already know the background and techniques of the immediate research well. Such knowledge allows the reader to move directly to the results and discussion.

Logical Development

The format of an empirical research report closely parallels the scientific method. Scientists typically start with questions for which they want answers. They then review the literature to see if other scientists have asked the same or similar questions. If the specific questions have not been answered, the scientists devise methods to get the answers, often by revising previously reported research. When the results of the research are in, scientists analyze them in the light of previous research to see what they mean and to arrive at conclusions. Because the overall format of the research report follows their research patterns so well, scientists find it easy to organize their research reports in this way.

Useful Elements

Scientists learn early in their studies the function and content of each element of a research report. They learn, for example, that when fellow scientists have finished reading the introduction, they should know the subject, scope, significance, and objective of the research. Knowing beforehand what each element of the report requires, report writers are far more likely to write useful reports for their fellow scientists than if they had to figure out each element anew.

A Caution

Notice the repeated use of the term *fellow scientists* in the preceding paragraph. The term illustrates a potential weakness in the empirical research report format and other formats, as well. Scientists get so accustomed to reporting their research in this format that they use it even when it's inappropriate for their audience and purpose.

Executives of many companies have difficulty getting their scientists to report research in ways that the executives can use it. Company management generally wants to know if reported research has commercial potential for the company. Will it, for example, lead to new, profitable products? Unless the company scientists report their research in ways that executives can understand and use for company decision making, the reports will not be useful.

As you will see in Chapter 10, many formats comparable to the empirical research report are in use. All help accessibility, selective reading, comprehension, and logical development. Use them wisely, not blindly. When they are not appropriate to your purpose and audience, modify them to meet your needs. As was said many times in Part One, know your purpose and your audience, and organize your reports around them.

9

Elements of Reports

This chapter describes every element you *might* need in a report, from the title page to the reference list. How many of these elements you actually use depends on the type of report you are writing and the needs of your audience. A large-scale company report might need every one of the elements. A short, informal, intraoffice report might need only a title page, introduction, discussion, and summary. Chapter 10 recommends the elements needed for specific kinds of reports, such as feasibility reports. Remember that the aim of all these elements, as with all format requirements, is to improve accessibility, selective reading, and comprehension for your readers.

Title Page

The *title page* will likely be the first thing a reader examines. It should, therefore, be useful and attractive.

To be useful, the title page should contain, at a minimum, the following information: the title of the report, the name of the author(s), the name of the person(s) for whom the report is written, and the date of the report. If those people writing and receiving the report have titles or organizational affiliations, include them, as well. Figure 9.1 shows a basic title page.

Antarctic Bottom Water Formation in the Northwestern Weddell Sea

Prepared for

Winifred Reuning

Editor

Antarctic Journal of the United States

by

Theodore D. Foster

Marine Sciences

University of California
Santa Cruz, California 95064

15 February 1993

FIGURE 9.1   Basic Title Page

Source: Adapted from Theodore D. Foster, "Antarctic Bottom Water Formation in the Northwestern Weddell Sea," *Antarctic Journal—Review* 27 (1993): 75–76.

In professional situations, the title page may also include such items as contract numbers, security codes, logos, and abstracts. Generally speaking, in such situations, you will be furnished instructions or samples to work from. If in doubt, ask for help from experienced people in your organization.

The wording of the title is important. Be sure it's complete enough to make your subject clear, but do not add useless words. The title "Volcanic Gases Create Air Pollution on the Island of Hawai`i" is good because it makes clear the subject of the report and contains nothing superfluous. A title such as "Gases Create Pollution in Hawai`i" would not truly identify the subject and would almost certainly mislead the reader into thinking the subject is broader than it is. On the other hand, adding such phrases as "A Report Concerning" or "A Study of" provides no useful information.

Make your title pages attractive by keeping them uncluttered and well balanced. Don't let word-processing capabilities tempt you into using a hodgepodge of type styles and excessive ornamentation. The judicious use of some boldface type and, when appropriate, a logo will generally suffice.

Letter of Transmittal

Most technical reports go to one person or to a small group of persons. Often, a *letter of transmittal* is used to formalize the forwarding of the report. When used, the letter of transmittal gives the following basic information: a statement of transmittal, the reason for the report, and the subject and purpose of the report. If you think it's appropriate, you may also point out special features of the report (such as specially prepared charts and graphs) and acknowledge people or organizations who have been helpful in preparing it. See Figure 9.2 for an example of a typical letter of transmittal.

In some circumstances, you may point out certain implications of the report and even state your major conclusions and recommendations. How much you include in your letter of transmittal depends to some extent on whether your report includes such features as executive summaries or informative abstracts (features that will be explained shortly). To aid in reader accessibility and selectivity, a certain amount of redundancy is permissible and even desirable in report format, but don't overdo it.

Marine Sciences
University of California
Santa Cruz, CA 95064
15 February 1993

Winifred Reuning, Editor
Antarctic Journal of the United States
National Science Foundation
Office of Polar Programs
4201 Wilson Boulevard
Arlington, VA 22230

Dear Ms. Reuning:

I submit the accompanying report, "Antarctic Bottom Water Formation in the Northwestern Weddell Sea," in accordance with the requirements of NSF Grant OPP 89-15730.

The report summarizes and analyzes results of the hydrographic study of Weddell Sea bottom water carried out from October through December in 1992. The study sampled for salinity, oxygen, nitrate, phosphate, silicate, and chlorophyll.

We have uncovered evidence that bottom water forms all year long in the western Weddell Sea; however, we will need further research to prove this conjecture.

Sincerely,

Theodore D. Foster

Theodore D. Foster

FIGURE 9.2   Letter of Transmittal

Source: Adapted from Theodore D. Foster, "Antarctic Bottom Water Formation in the Northwestern Weddell Sea," *Antarctic Journal—Review* 27 (1993): 75–76.

Depending or your organization's policy, the letter of transmittal may be placed immediately before or after the title page. Alternatively, the letter of transmittal may be mailed separately as notice that the report is forthcoming.

Preface

Sometimes, reports are intended for large groups of people. For example, environmental impact statements are used to explain the environmental impacts of building large projects, such as highways. Because many people are concerned with such impacts, such reports are widely circulated. When this is the case, a *preface* is more appropriate than a letter of transmittal. A preface differs little from a letter of transmittal except in format (see Figure 9.3).

Preface

This report has been prepared by the U.S. Geological Survey's National Landslide Information Center primarily for the Colorado Department of Transportation (CDOT), the city government of Golden, Colorado, and the citizens of Golden. During 1993, landslide conditions closed the southbound lane of the highway bypass around Golden. Despite repeated attempts to stop the slide's downslope movement, it continues to encroach on the highway.

Causes of the landslide include steep slopes, heavy precipitation, and the presence of soils and bedrock susceptible to landslides. The presence of all these factors means there are no easy, inexpensive solutions to the problem. This report examines the problem in detail and offers solutions to the problem for the consideration of the CDOT and the citizens and officials of Golden.

We thank the CDOT for furnishing us CDOT research results from its extensive investigations into the problem.

FIGURE 9.3   Preface

Source: Adapted from Lynn M. Highland and William M. Brown III, "The Golden Bypass Landslide, Golden, Colorado," *Earthquakes & Volcanoes* 24 (1993): 4–14.

Table of Contents

A *table of contents* (*TOC*) aids accessibility and selectivity by locating the major divisions and subdivisions within your report for readers. All division and subdivision headings in your TOC must match, word for word, the corresponding headings within the report. You may have more than three levels of headings within your report (see pp. 32–38), but generally, it's not practical for the TOC to have more than three levels. If you make your TOC overcomplicated, it will quickly become as difficult to find things in the TOC as in the report.

In constructing your TOC, use some combination of capital letters, boldface, and indentations to make the different levels of headings dis-

FIGURE 9.4   Typical Table of Contents

Source: Adapted from Isobel Schell, "Counternarcotics: Prospects for Success," *Geographic and Global Issues Quarterly* Spring 1994: 6–9.

tinct from one another. As with the title page, however, don't be tempted into excessive complication by your word-processing capabilities. Figure 9.4 shows a typical TOC, suitable for a student report. Figure 9.5 shows a professionally produced TOC. You can find many other examples in the books, magazines, and journals that you read.

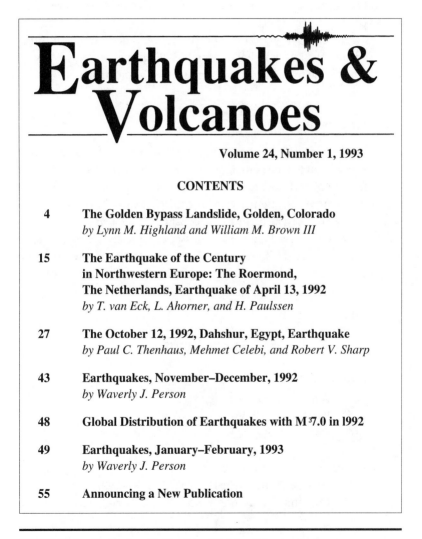

Earthquakes & Volcanoes

Volume 24, Number 1, 1993

CONTENTS

FIGURE 9.5   **Professionally Produced Table of Contents**

Source: Earthquakes & Volcanoes 24 (1993).

List of Illustrations

If you have more than three or four figures and tables in your report, you may want to include a listing of them immediately after the TOC. If you mix tables and figures in one list, call it a *List of Illustrations* (see Figure 9.6). If you have separate lists, call them *List of Tables* and *List of Figures*. You can be even more specific. For example, if you have many maps, create a *List of Maps*.

As with report titles, illustration titles should describe the illustrations adequately but not include useless language, such as "A Map Showing . . ." The reader can see it's a map. (See also Chapter 6.)

Glossary

Analyze your audience. Are you using words that are unfamiliar to your readers? If so, you will need to define those words. If you have only a few to define, you may choose to do so in the report—perhaps in the introduction or where you use the word for the first time. If you use many unfamiliar words, say, as many as 10, consider using a *glossary*—that is, a list of definitions (see Figure 9.7). Glossaries typically use parallel sentence fragments as definitions, most often noun phrases. However, if the definitions are extended past the fragments, complete sentences are

---

ILLUSTRATIONS

MAPS
1. Active Volcanoes in Alaska
2. Westdahl Peak Volcano
3. Cook Inlet Region

TABLES
  I. Alaskan Earthquakes per Week, 1991–1992
 II. Californian Earthquakes per Week, 1991–1992

FIGURE 9.6   List of Illustrations

GLOSSARY

Ash. Rock fragments (diameter < 2 mm) created by explosive volcanic eruptions.

Caldera. A huge, nearly circular volcanic depression created by the explosive eruption and collapse of a volcano.

Cinder Cone. A steep, cone-shaped hill formed by the accumulation of cinders and other material ejected from a volcanic vent.

Lahar. A flowing mixture of water and rock debris, sometimes referred to as debris flows or mudflows, that forms on the slopes of a volcano.

Lava. Molten rock extruded from a volcano or volcanic fissure.

Lava Flow. Molten rock that flows from a volcanic vent.

Lava Fountain. Jets of incandescent lava from a volcanic vent or fissure.

Magma. Molten rock within the Earth containing gas and crystals.

Pyroclastic. Pertaining to fragments of lava or rock ejected explosively from a volcanic vent.

Pyroclastic Flow. A hot, fast-moving, avalanche of gas-charged pyroclastic debris such as ash, pumice, and rock fragments discharged during explosive eruptions.

Tephra. A general term for airborne rock fragments ejected during explosive eruptions. Tephra consisting of fragments less than 2 mm in diameter is called ash.

Volcanic Tremor. A type of seismicity characterized by continuous vibration of the ground related to the transport of fluids and gas within or beneath a volcano.

FIGURE 9.7   Glossary

Source: Game McGimsey, "Volcanic Activity in Alaska: September 1991–September 1992," *Earthquakes & Volcanoes* 24 (1993): 73.

used. The entry for *Tephra* in Figure 9.7 makes the concept clear. (Also see Chapter 2 for more about definitions.)

You may locate the glossary in the front of your report, probably after the TOC or the list of illustrations, or in the back as an appendix.

Be sure to state in your introduction where the glossary is. Some report writers go so far as to use boldface or a symbol of some sort (such as an asterisk) to identify in the text words that are defined in the glossary.

List of Symbols

Scientific and technical writing often includes a good many symbols, most of which need definition. As with words, you can define symbols in the text of your report or in a separate list (see Figure 9.8). The *list of symbols* usually is located in the front of the report, following the list of illustrations or the glossary.

Abstracts and Summaries

The readers of technical reports are busy people. They need to have the key points of the reports they read summarized for them (see Chapters 2 and 3). *Abstracts* and *summaries* serve that purpose. Each condenses

---

LIST OF SYMBOLS

TS   Target strength in decibels

s   Acoustic cross section

l   Wavelength of the acoustic signal in centimeters

ETL Estimated target length in meters

s_{bs} Weight specific backscattering cross section in square meters per milligram

WW   Krill wet weight in milligrams

FIGURE 9.8   List of Symbols

the most important facts, generalizations, conclusions, and recommendations of a report into a concise statement. Whether you call that statement an *abstract* or a *summary* depends mainly on the kind of report you are writing and where you locate the statement.

Many scientific and technical reports have a summarizing statement near the front. In that position, the statement is most often called an *abstract*. If placed at the end of the report, the statement is usually called a *summary*. Reports written specifically for executives usually contain a special kind of summary called an *executive summary*, most often placed just before the introduction. The sample report formats in Chapter 10 make clear where summaries and abstracts are placed in various kinds of reports.

Figure 9.9 is a summary found at the end of an article written for a mixed audience of generalists and specialists. It concisely restates the key facts and generalizations of the article.

Summary

In comparison with worldwide contributions from anthropogenic sources, such as power plants and industrial processes, Kilauea Volcano contributes an insignificant amount of sulfur dioxide and carbon dioxide to the atmosphere. However, Kilauea is a significant local source of air pollution on the island of Hawai`i.

Since the beginning of nearly continuous activity on Kilauea in 1986, vog [volcanic smog] has been reported not only on the island of Hawai`i but also during certain wind and weather conditions, along the entire 500-km-long populated length of the island chain. Although our ability to mitigate the effects of volcanic air pollution are limited, regional burning bans on the island of Hawai`i is instituted during vog episodes to prevent further deterioration of air quality. The present level of volcanic air pollution on the island of Hawai`i will probably persist as long as the current steady effusion of lava and gas continues.

FIGURE 9.9   Summary

Source: Jeff Sutton and Tamar Elias, "Volcanic Gases Create Air Pollution on the Island of Hawai`i," *Earthquakes & Volcanoes* 24 (1993): 195.

Figure 9.10 is an abstract that appears before a report of scientific research. Like all abstracts, it's meant to stand alone, if need be. Therefore, it contains the key facts, generalizations, and conclusions of the report.

Figure 9.11 is an executive summary. Like many, it defines a problem, considers a solution, and ends with a recommendation.

A *descriptive abstract* is different from other abstracts and summaries. As its name implies, it describes the report. That is, rather than summarizing the report, it briefly tells what will be found in the report,

ABSTRACT

The customary approach to classifying multiple audiences for written discourse is to recognize primary, secondary, and immediate audiences, and, in some cases, gatekeeping audiences. Based on findings from an ethnographic case study of engineering authors in an R&D setting, this article suggests that authors should also attend to *watchdog* audiences as they write. A watchdog audience pays close attention to the written transaction between the author and the primary audience. Authors must direct their discourse toward the primary audience, but they must also keep the motives and purposes of the watchdog audience in mind as they write and revise. The watchdog audience in my case study, while it had no direct leverage or other organizational power over the authors, still influenced the authors extensively as they revised their text. Evidence indicates that, beyond the apparent and traditional sources of power, there are more contextual, hidden, socially mediated power relationships equally capable of shaping written discourse.

FIGURE 9.10   Abstract

Source: Vincent J. Brown, "Facing Multiple Audiences in Engineering and R&D Writing: The Social Contexts of a Technical Report," *Journal of Technical Writing and Communication* 24 (1994): 67.

SUMMARY

Despite some law enforcement successes against the global drug-trafficking industry, narcotraffickers have adapted successfully to mounting law enforcement efforts. Insufficient political will, widespread governmental corruption, and the enormous profits involved all work against law enforcement efforts. In addition, the traffickers' vast resources allow them to use state-of-the-art technology, such as multiengined aircraft and containerized shipping, to thwart interdiction efforts. A counternarcotics strategy that emphasizes eliminating drug crops may offer some advantages over law enforcement strategy. Elimination can be achieved through crop eradication and the development of alternative crops.

Crop eradication: Aerial eradication by herbicides has long been recognized as cost effective and safe. Because the coca bush takes 1.5 years to become productive, unimpeded eradication could destroy a large portion of the crop in one year. Though aerial eradication faces stiff political opposition in Peru and Colombia, changing political conditions may make it more feasible.

Alternative crops: Coca growers have long indicated willingness to grow other crops if they were economically feasible. Because of economic reforms allowing for better marketing of Latin American crops, the encouragement of alternative crops to replace coca in Latin America seems possible. US-backed alternative development programs have already increased land use devoted to noncoca crops in Bolivia.

Law-enforcement measures without eradication and alternative crop programs will have only limited success. In Latin America, the rapid move toward economic liberalization, privatization of industries, and relaxation of trade barriers could provide opportunities for ending dependency on the illicit drug trade. We should take advantage of the opportunities presented by these changes to encourage and assist the governments involved to develop eradication and alternative crop programs.

FIGURE 9.11   Executive Summary

Source: Adapted from Isabel Schell, "Counternarcotics: Prospects for Success," *Geographic and Global Issues Quarterly* Spring 1994: 6–9.

thereby prompting readers to decide whether they want to read the report that follows. In a company report, the descriptive abstract is often placed on the title page. In a journal article, it may be placed above or below the article title, serving much like an extended title. What follows is typical:

> This article describes the design and evaluation of a formal writing assessment program within a technical writing course. Our purpose in this baseline study was to evaluate student writing at the conclusion of the course. In implementing this evaluation, we addressed fundamental issues of sound assessment: reliability and validity. Our program may encourage others seeking to assess educational outcomes in technical writing courses.[1]

Introduction

Before reading what follows, read the *introduction* in Figure 9.12. Have it in mind as you read this section.

An introduction *must do* these two things:

- Announce the subject of the report.
- Announce the purpose of the report.

An introduction *may do* any of these four things:

- Catch the reader's interest in the article or report.
- Define terms and concepts.
- Provide theoretical and historical background.
- Forecast the content and organization of the report.

How many of the optional things your introduction does depends on your audience and plan for your report or article. For a general, non-technical audience, you would likely begin with something to catch the reader's interest. If your report or article uses words and concepts that the audience will need defined and explained to understand your

Introduction

In a handful of molten magma weighing about a pound, there is less than a tenth of an ounce, by weight, of dissolved gas—roughly the same weight as a pinch of table salt. Yet this tiny amount of gas produces spectacular lava fountains hundreds of meters high. The fountaining occurs as magma reaches the surface, because dissolved volcanic gases exsolve and expand tremendously as pressure on the magma is released. Anyone who has shaken a bottle of soda and opened it quickly has received the full value of this basic principle of physics.

Gases are dissolved in magma at depth, where pressures within the Earth's crust are very great—many thousands of pounds per square inch. As the magma rises to the surface and erupts, the pressure decreases, and gas is released. The main gases dissolved in magma are water vapor, carbon dioxide, and sulfur gases, with lesser amounts of others, such as hydrogen, carbon monoxide, hydrochloric acid, and hydrofluoric acid. In our pinch-of-salt-to-a-handful-of-magma illustration above, most of the "pinch" is water vapor, followed by lesser amounts of carbon dioxide and sulfur gases with a few "grains" of hydrogen and the other acid gases.

The current eruption of Kilauea produces large quantities of volcanic gases that contribute to "volcanic air pollution." In this article we discuss the nature of the gases released from Kilauea, how we study them, and what happens to the gases in the environment after they are released.

FIGURE 9.12   Introduction to Journal Article

Source: Jeff Sutton and Tamar Elias, "Volcanic Gases Create Air Pollution on the Island of Hawai`i," *Earthquakes & Volcanoes* 24 (1993): 178.

subject matter, you may choose to provide the definitions and explanations in the introduction. However, you may also choose to provide them elsewhere in the article, perhaps where you use each word or concept. The same holds true for providing needed theoretical and historical background.

If your report or article is complex, you would be wise to forecast the content and organization. For a short, uncomplicated report or article, you could probably forego this step, but it would always be appropriate to include it.

No matter what else you *may* do in your introduction, you *must* announce your subject and purpose. In other words, tell your readers what you are talking about and why.

The excellent introduction in Figure 9.12 introduces an article in a publication with a mixed audience. The publication's purpose, as announced in its masthead, is "to provide current information on earthquakes and seismology, volcanoes, and related natural hazards of interest to both generalized and specialized readers."[2] The first two paragraphs of the introduction provide theoretical background and define concepts. The first sentence is an interest catcher because of the disparity between the amount of magma and the amount of dissolved gas. The first two paragraphs provide information that will be appreciated by the general audience, even if it's unneeded by the specialists. The third paragraph makes the subject and purpose clear and forecasts the content and organization of the article.

Introductions to articles aimed squarely at general audiences tend to emphasize interest catching and to be more informal than introductions aimed at more specialized audiences. In these general-audience introductions, the subject and purpose will be announced but perhaps somewhat indirectly. Figure 9.13 illustrates such an introduction. It emphasizes human interest to catch the reader's attention. Some inference is required of the reader, but it's reasonably clear that the subject will be the new food labels required by the government and the purpose will be to show the reader how to use these labels to choose "heart-healthy" foods. Creating human interest, as is done in Figure 9.13, is appropriate, but don't get too breathless about it.

Articles aimed exclusively at specialized audiences seldom bother with interest-catching introductions. They tend to get right down to business and make the subject, purpose, content, and organization quite clear. The introduction in Figure 9.14 well illustrates such an introduction.

Remember, introductions vary, depending on function and audience, but an introduction that doesn't announce the subject and purpose is incomplete.

My mother, an on-again, off-again low-fat, low-cholesterol dieter, rushed up to me in the grocery store one day last year. She was clutching a package of turkey frankfurters. Knowing I'm a registered dietitian, she pointed to the 5 milligrams of cholesterol listed on the package's nutrition panel and said, "Now, tell me: Is this high or low in cholesterol?"

If she had been holding a package with the new Nutrition Facts panel, I wouldn't have had to stand there sputtering and stammering as I did, waiting for the answer to come to me. Instead, I would have quickly referred her to the % Daily Value column on the panel's right side.

There, she would have seen at a glance that a serving of those turkey franks (two of them, about 55 grams) provided only 2 percent of the Daily Value for cholesterol. As a rule of thumb, foods containing 5 percent or less of the Daily Value for a nutrient are low in that nutrient. So, a serving of those franks was low in cholesterol.

Now when my mother and others like her shop for "heart-healthy" foods, they can easily find this information on many products. Regulations requiring it and other labeling changes went into effect for many food products May 1994, and many now carry the new label.

The regulations come from the Food and Drug Administration and the U.S. Department of Agriculture. FDA's rules implement the Nutrition Labeling and Education Act of 1990.

FIGURE 9.13   Introduction for a General Audience

Source: Paula Kurtzweil, "The New Food Label: Help in Preventing Heart Disease," *FDA Consumer* December 1994: 19–20.

Discussion

The *discussion,* where you fulfill your subject and purpose, will be the longest part of your article or report. How you organize and write it will depend on your purpose and audience. For direction, see Chapters 1 through 7 and remember the seven principles of technical writing:

1. Know your purpose.
2. Know your audience.
3. Choose and organize your content around your purpose and audience.

This study presents the results of a two-year naturalistic case study of how 6 children in two different instructional settings acquired alphabetic knowledge as they developed as readers and writers. In this study, alphabetic knowledge included knowledge of the graphemic and phonemic nature of written language, grapheme/phoneme correspondences, and use of graphophonics as a tool for reading and writing. This longitudinal study examined these children's development as it occurred in both skills-based and whole language classrooms. We begin by briefly framing this study within the historical debate on phonics instruction and previous research on children's development of alphabetic knowledge. After describing our approach and results, we discuss what our study may mean to the fields of beginning literacy development and instruction.

FIGURE 9.14   Introduction for a Specialized Audience

Source: Ellen McIntyre and Penny A. Freppon, "A Comparison of Children's Development of Alphabetic Knowledge in a Skills-Based and a Whole Language Classroom," *Research in the Teaching of English* 28 (1994): 391–92.

4. Write clearly and precisely.
5. Use good page design.
6. Think visually.
7. Write ethically.

Conclusions and Recommendations

Analytical reports require *conclusions*. Recommendation reports require *conclusions* and *recommendations*. *Conclusions* are opinions based on the data in reports. For example, a recent journal article analyzed data that indicate that ulcers are caused by bacteria—not stress, as previously thought. The last paragraph of the article expressed a conclusion based on the data:

Meanwhile, the future of current ulcer sufferers looks brighter than ever. Says consensus team member Ann L. B. Williams, M.D., of George Washington University Medical College, "We now have an opportunity to cure a disease that previously we had only been able to suppress or control."[3]

Whereas *conclusions* are opinions based on the data presented, *recommendations* are the actions recommended (or recommended against) based on the conclusions. Recommendation reports, often called *feasibility reports*, are common in business and government organizations. The studies that lead to such reports examine problems. For example, researchers working for a state prison system may be asked to look for ways to reduce the increasing pressure on prison facilities caused by growing prison populations. After due study, the researchers may reach a conclusion such as this one:

> A major conclusion, based on our data, is that society and nonviolent offenders might be better served by having such offenders perform community service over several years rather than serve prison terms. Removing nonviolent offenders from the prison population would help reduce the overcrowding that exists.

Based on such a conclusion, the researchers then make a recommendation, such as this:

> We recommend legislation allowing and encouraging judges to sentence nonviolent offenders to long-term community service, rather than prison. The legislation should authorize follow-up research to analyze the value of the community service and its effect on the recidivism rate of those sentenced to such service.

Because conclusions are opinions and because recommendations are based on conclusions, be sure the conclusions you reach are well grounded on reliable data.

Appendixes

As their name indicates, *appendixes* are items appended to the main body of a report. They are excellent devices to help satisfy the needs of a dual audience for a report. For example, suppose the prison feasibility study described in the preceding section has two audiences: legislators and legislative aides. The legislators will want to know the salient facts and the conclusions and recommendations reached. They will not want to be buried under accounts of research methods and the like. However, the legislative aides may need such information to evaluate the study.

Putting the detailed information in an appendix makes it available for the aides but out the way of the legislators.

You can make two mistakes in selecting material for an appendix. You may segregate material in the appendix that everyone in your audience needs and wants and, therefore, run the risk of it being overlooked. Conversely, you may load your appendix with material that nobody needs or wants, increasing the bulk of your report but not its value. As always, let your audience and purpose guide you in reaching such decisions.

Documentation

Most technical reports require *documentation:* the use of references to identify material you relied on in preparing the report. References credit your sources and allow your readers to find them, if they wish. Many documentation systems exist. If you are preparing an article for a journal, you need to obtain the style book used by that journal as a guide to its documentation system. Likewise, companies, government agencies, university departments, and so forth may all require some special systems of documentation.

It all seems a bit bewildering, but most documentation systems require the same information: author's name, editor's name (if any), title of book or article, and publication data. Publication data in the case of a book include the publisher's name, city of publication, and date of publication. When necessary, publication data may also include such information as edition numbers, volume numbers, and series numbers. Publication data for an article would include the page numbers of the article and the name, volume, number, and date of the periodical.

The differences in documenataion systems involve mainly differences in punctuation, capitalization, and the order in which information is presented. The best way to learn a system is to obtain the style book involved and, when you are documenting your report, imitate the appropriate formats down to the last period. Meanwhile, pay attention to what you are seeing and doing, particularly to punctuation, capitalization, and order.

Provided here are samples based on *The Chicago Manual of Style's* author-date system. It is used in many of the natural and physical sciences and in some of the social sciences. It is, therefore, a common system. It is also, as documentation systems go, a fairly simple one.[4] You will find here enough samples to see you through a typical report. If you need more than is provided, see *The Manual* itself, readily available in most libraries.

Documentation using the author-date system requires adding author-date references in the text that refer the reader to an alphabetized list called by such titles as *References* or *Works Cited.* Place each author-date reference within parentheses in the text, as illustrated in Figure 9.15. Their actual format depends on the information you have to provide, as illustrated here. As you use these samples, carefully note punctuation, capitalization, and order:

Basic Format

(Baker 1992)

Reference to Specific Page or Division

(Baker 1992, 74)
(Baker 1992, Ch. 9)

Reference to Volume

(Cornwall 1993, vol. 2)

Reference to Volume and Page

(Cornwall 1993, 2:67)

Two or Three Authors

(Fielding and Meaders 1989)
(Manchester, Kehoe, and Holl 1994)

Most teachers who have used students to tutor other students have concluded that students given such experience learn more than those who have not been given such opportunities (Cohen, Kulik, and Kulik 1982; Cohen and Riel 1989; Duin and others 1994). As Bruffe recognized, "The tutors' commitment to the value of their own and each other's writing and the quality of their writing improved dramatically" (1978, 451).

FIGURE 9.15   Author-Date References within Text

More Than Three Authors

(Osborn and others 1993)

Author with Two or More Works of Same Date Cited

(Larsen, Generic drugs, 1992)
(Larsen, Generic painkillers, 1994)

Organization as Author

(Landings Corporation 1995)

Multiple References in Same Parentheses

(Baker 1992; Lewis 1993; Noelani 1995)

Author's Name Used in Text

(1992)
(1992, 74)

Note: When you cite authors directly in text, do not repeat their names in parenthetical references (see Figure 9.15).

Figure 9.16 illustrates how to construct an alphabetized reference list. Here are identified typical entries for such a list:

Basic Book

Rock, I. 1975. *An introduction to perception.* New York: Macmillan.

Note: This entry lists, in order, the author, date, title, city of publication, and publisher. You may use initials or full names of authors, but be consistent throughout the reference list. The names of publishing companies are usually given in short forms. For example, the *Macmillan Publishing Company* is listed as *Macmillan.*

Book with Two or More Authors

Spoehr, K. T., and S. W. Lehmkuble. 1982. *Visual information processing.* San Francisco: W. H. Freeman.

Note: Do not use "and others" in a reference list. List all the authors—last name first for the first author, and normal order for the rest.

REFERENCES

Coe, R. 1992. *Process, form, and substance.* 2d ed. Englewood Cliffs, NJ: Prentice Hall.

Colomb, G. G., and J. Simutis. 1992. Written conversation and the transition to college. Paper presented at the Computers and Writing Conference, May, at Ann Arbor, MI.

Cooper, M. 1989. The ecology of writing. In *Writing as social action,* ed. M. M. Cooper and M. Holzman. 1–13. Portsmouth, NH: Boynton Cook/Heinemann.

Cunningham, D. Letter to author, 28 October 1995.

Eisner, E. W. 1985. *The educational imagination.* 2d ed. New York: Macmillan.

— — —. 1991. *The enlightened eye.* New York: Macmillan.

Fahnestock, J. 1993. Genre and rhetorical craft. *Research in the Teaching of English* 27: 265–71.

Frank, F. W., and P. A. Treichler, eds. 1989. *Language, gender, and professional writing.* New York: MLA.

Rock, I. 1975. *An introduction to perception.* New York: Macmillan.

FIGURE 9.16   Reference List

Book with Editor

Frank, F. W. and P. A. Treichler, eds. 1989. *Language, gender, and professional writing.* New York: MLA.

Note: Use *ed.* for editor, *eds.* for two or more editors, and *trans.* for singular or plural translators. Names of organizations likely to be known to readers are often abbreviated—in this case, *MLA* for the *Modern Language Association.*

Organization as Author

United States Environmental Protection Agency. 1991. *Citizen's guide to insecticides.* Washington, DC: GPO.

Note: GPO is a widely used abbreviation for the *Government Printing Office,* which prints most books published by the federal government.

Later Editions

Coe, R. 1992. *Process, form, and substance.* 2d ed. Englewood
Cliffs, NJ: Prentice Hall.

Note: When the city of publication is not well known, use additional identification, such as the state abbreviation.

Essay in an Edited Collection

Cooper, M. 1989. The ecology of writing. In *Writing as social
action,* ed. M. M. Cooper and M. Holzman. 1–13.
Portsmouth, NH: Boynton/Cook Heinemann.

Basic Journal Entry

Fahnestock, J. 1993. Genre and rhetorical craft. *Research in the
Teaching of English* 27: 265–71.

Note: This entry lists, in order, the author, date, article, journal, volume, and inclusive pages for the article. Use this form for journals
that paginate by volume, rather than issue. Treat the names of multiple authors for articles as you do multiple authors for books. List
page numbers as in these examples:

1–13; 16–28; 200–206; 201–8; 224–29; 1156–68.

Entry for Journal That Paginates
by Issue

Farley, D. 1994. In-home tests make health care easier. *FDA
Consumer* 10 (December): 25–28.

Note: Omit the volume number. Use the issue number and the date
in the journal masthead—for example, *December, Fall, June 18,* and
so forth.

Paper Read at a Meeting

Colomb, G. G., and J. Simutis. 1992. Written conversation and
the transition to college. Paper presented at the
Computers and Writing Conference, May, at Ann Arbor,
MI.

Personal Communication

Cunningham, D. Letter to author, 28 October 1995.

Note: Identify the nature of the communication: letter, telephone call, interview, and so forth.

Two Entries for an Author

Eisner, E. W. 1985. *The educational imagination.* 2d ed. New York: Macmillan.

— — —. 1991. *The enlightened eye.* New York: Macmillan.

Note: Use three long dashes (or six hyphens) for another entry by the same author.

For examples of the Modern Language Association (MLA) documentation style, frequently used in the humanities, see the chapter notes in this book.

Copyright

Copyright laws protect most published work. Exceptions are most materials published by the United States government or state and local governments. Such work is normally not copyrighted.

If your work will be unpublished—for example, a student report— you must identify your sources, but you do not need permission to use copyrighted material. In general, if you plan to publish your work, you must get permission from the copyright holders (usually, the publishers) to use figures and extended quotations from copyrighted work. See *The Chicago Manual of Style* concerning copyright law.

ENDNOTES

[1]Norbert Elliot, Margaret Kilduff, and Robert Lynch, "The Assessment of Technical Writing: A Case Study," *Journal of Technical Writing and Communication* 24 (1994): 19

[2]*Earthquakes & Volcanoes.*

[3]Ricki Lewis, "Surprise Cause of Gastritis Revolutionizes Ulcer Treatment," *FDA Consumer* December 1994: 18.

[4]*The Chicago Manual of Style,* 14th ed. (Chicago: The University of Chicago Press, 1993).

10

Formats of Reports

Business and professional people write an assortment of reports. In these reports, they instruct, analyze information, propose work to be done, report progress on work, and report and interpret the results of research. This chapter describes how to format this variety of reports.

Instructions

Most instructions have a basic three-part format: an introduction, a list of equipment and materials needed, and how-to instructions. If you have ever followed a recipe in cooking, the format will be immediately familiar to you. Many instructions also include warnings, theory, and a glossary.

Introduction

An introduction to instructions must announce the subject and purpose of the instructions, something like this:

> Leaking faucets waste water, stain the sink, and create annoying dripping sounds. With the right equipment, stopping the leak is a simple process.

This simple introduction makes clear that the subject is *stopping a faucet leak* and the purpose is to show you how to do it. In addition, this introduction provides motivation for doing the task—stop waste, staining, and annoying sounds.

If the process you are going to describe is complex, you might also preview it:

> The process involves turning off the water to the faucet, disassembling the faucet, replacing the faucet washer, reassembling the faucet, and turning the water on again.

You may also include such things as warnings, references to a glossary, and definitions, but in general, keep introductions uncomplicated.

List of Equipment and Materials

People about to do a task need to know what they must have to complete the task. A list detailing the needed equipment and materials provides the necessary information:

- A box of washers of assorted sizes
- A screwdriver
- An adjustable wrench

How much information you provide in your list depends on your audience analysis. If your analysis tells you that your readers are experienced tool users, the simple list above should be sufficient. If you have inexperienced tool users, you may have to expand the list:

- A screwdriver of the appropriate type and size. Screws have either straight-blade slots or phillips slots (see Figure 1). The screwdriver must match the slot. In either case, the screwdriver blade must fit securely into the slot of the screw without slippage (see Figure 2).

As this expanded list indicates, figures (not shown here) make clear the distinction between a straight-blade slot and a phillips slot and how the screwdriver blade should fit the slot. As Chapter 6, Think Visually, emphasizes, when pictures are clearer than words, use pictures.

Sometimes, you may find that your readers need a rather complete description of a tool or mechanism involved in a process. In describing mechanisms or tools, you can include information on their purpose and function, parts and subparts, purpose and function of the parts and subparts, construction, materials, appearance, size, color, and so forth. You can also describe how to use the mechanism safely. How much of this sort of information you include depends, as always, on what your readers actually need. Figure 10.1 provides a good example of such a description. Again, notice the use of illustrations in the description.

If you think it's necessary, you can also include information about where materials and equipment can be obtained, what they cost, and so forth. Tell your readers what they need to know to do the job properly.

How-To Instructions

Figure 10.2 illustrates a partial list of how-to instructions. Look at it now, before you continue with the text. As Figure 10.2 demonstrates, how-to instructions follow these principles:

- Each instruction of process presented separately and in chronological order
- Use of simple language, active voice, and imperative mood: "Loosen packing nut with wrench"
- Use of clarifying illustrations
- Inclusion of helpful advice: "Most nuts loosen by turning counter-clockwise"

Although how-to instructions are not always numbered (as in Figure 10.2), they frequently are.

It's critically important in writing how-to instructions to break the process down into manageable instructions. An instruction may include only one step (as in instruction 3 in Figure 10.2) or several (as in instructions 1 and 2). When you include more than one step in an instruction, be sure they are all closely related.

If you are experienced in the process being described, you may unintentionally leave out steps that you have come to do almost automatically, without thinking about them. Be sure to think through the entire

Portable Power Circular Saws

The portable power circular saw can save you "muscle power" and time (fig. 12). You can rent or buy one. It may be used as a crosscut saw or a ripsaw—depending upon the type of blade used.

The *saw blade* should be adjusted so that the amount of blade that extends below the "shoe" is slightly greater (1/16 to 1/8 inch) than the thickness of material to be cut. As you guide the saw forward, the blade is exposed for cutting (fig. 13).

For ripping work, circular saws come with a "ripping guide." After adjusting the blade, set the ripping guide the same distance from the saw as the width of the material to be cut off.

Then place the guide against the edge of the piece as you cut (fig. 14).

For crosscutting, or cutting off material, turn the ripping guide upside down, so that it will be out of the way. Using a framing square and pencil, draw a line to mark where to cut. Then guide the saw blade carefully along the line.

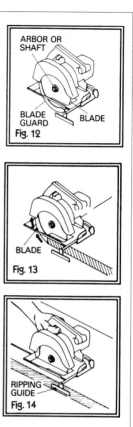

Using a portable power saw can save much time and effort. For safety and the proper use of the saw, follow these steps:

1. Make sure that the saw you use is equipped with a guard that will automatically adjust in use so that none of the teeth are exposed above the work.
2. Make sure the saw is equipped with an automatic power cutoff button.
3. Always wear goggles or face mask when using a power saw.
4. Carefully examine the material and make certain that it is free of nails or other metal before you start cutting.
5. Grasp the saw with both hands and hold it firmly against the work.
6. Never overload the saw motor by pushing too hard or cutting material that is too thick for this small saw.
7. Always try to make a straight cut to keep from binding the saw blade. If it does bind, back the saw out slowly and firmly in a straight line. As you continue with the cutting, adjust the direction of the cut so that you are cutting in a straight line.
8. Always pull the electric plug before you make any adjustments to the saw or inspect the blade.

FIGURE 10.1   Mechanism Description

Source: U.S. Department of Agriculture, *Simple Home Repairs: Outside* (Washington, DC: GPO, 1986) 6.

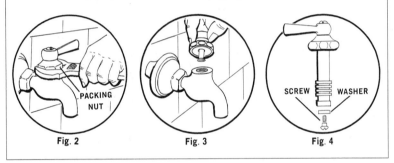

HOW-TO

1. First turn off the water at the shut-off valve nearest to the faucet you are going to repair. Then turn on the faucet until the water stops flowing. (Fig. 1)

2. Loosen packing nut with wrench. (Fig. 2) (Most nuts loosen by turning counter-clockwise.) Use the handle to pull out the valve unit. (Fig. 3)

3. Remove the screw holding the old washer at the bottom of the valve unit. (Fig. 4)

Fig. 1

VALVE

PACKING NUT

Fig. 2

Fig. 3

SCREW   WASHER

Fig. 4

FIGURE 10.2   How-To Instructions

Source: U.S. Department of Agriculture, *Simple Home Repairs: Inside* (Washington, DC: GPO, 1986) 1.

process, step by step; leave out nothing that your audience analysis tells you your reader needs. A good check is to have someone of the skill level you expect in your audience perform the process following your instructions. Gaps in your instructions will show up quickly when you do.

Warnings

To protect consumers from injury and to protect companies from expensive lawsuits, warnings are given extensively in instructions. They may stand out in a separate section of their own or be part of the introduction, list of equipment and materials, or how-to instructions. Manufacturers have learned (to their sorrow) that risks that seem obvious to them are not obvious to everyone. Err on the side of too many warnings, not on too few.

Make your warnings stand out so that no one will miss them. Box them; use a larger, distinctive, or different-colored typeface; use symbols, such as an exclamation point or a skull and crossbones. When the risk involved might lead to death or serious injury, use all three techniques.

Although there is not complete agreement on levels of warnings, three levels have become common: *Caution, Warning,* and *Danger:*

- Use *Caution* to warn against actions that may lead to undesirable results but that are not likely to damage equipment or injure people:

Caution

If you have external devices, such as an external hard drive, connected to your computer, turn them on before you turn on your computer. Failure to do so may keep your computer from recognizing them when you do turn them on.

- Use *Warning* to warn against actions that may damage equipment or materials or cause mild injury to users:

WARNING
This process will erase all information on this disk.

- Use *Danger* to warn against actions that may result in serious injury or death:

DANGER!
Do not stand on the top two steps or the top of this
stepladder. Doing so may result in a fall that
could cause serious injury or death.

Theory

Sometimes, users of instructions may benefit from knowing the theory that underlies the instructions. Knowing the theory may clarify the procedures described or motivate users to follow the instructions. In the following example, the author of instructions meant for managers of seedling nurseries explains the theory underlying the instructions:

> Managers of container seedling nurseries sow multiple seeds per cell to increase the probability of having at least one germinant per cell. This ensures that their greenhouses are fully stocked so that seedling contracts may be filled. However, this practice wastes valuable seed and necessitates thinning extra germinants at an additional cost. Multiple sowing, therefore, should be minimized.
>
> Many factors influence how many seeds per cell to sow, including species, seed size, seed availability and cost, type and accuracy of sowing equipment, sowing and thinning labor costs, and germination data reliability. The primary factor, however, is greenhouse germination percentage. When germination percentage is known or assumed, nursery managers use various rules of thumb or rely on the probability tables found in Tinus and McDonald (1979) to determine the number of expected empty cells. These tables are complete but sometimes cumbersome and currently unavailable to new managers.
>
> Fortunately, the percentage of empty cells can be obtained using a hand-held calculator (Schwartz 1993). Taking this procedure one step farther, the probability tables of Tinus and McDonald (1979), showing both filled and empty cells, can be recreated on a personal computer.[1]

If the theory you present is lengthy, you may want to put it in a section by itself. Most of the time, however, you will present theory as part of your introduction, as was the case in the previous example. Include theory in instructions only if you are reasonably sure your readers will benefit from it.

Glossary

Occasionally, you will use enough words unfamiliar to your readers to justify including a glossary with your instructions. When such is the case, construct your glossary following the instructions on pages 88–90. You may place your glossary at the beginning or the end of your instructions.

Analytical Reports

Analytical reports analyze data to arrive at conclusions. Some reports go one step further and recommend that actions be taken or not taken, based on the conclusions reached. If the person making the report has the authority to do so, the report might state a decision. Therefore, analytical reports may also be known as *recommendation reports* or *decision reports*. When the purpose of the report is to examine the feasibility of some plan of action, it may also be called a *feasibility report*.

The executives of organizations are constantly making decisions based on answers to questions such as these:

What can be done about the high absentee rate in our Charleston plant?

What is the sales potential of our new VCR?

What are the implications for the dollar of the U.S. federal government's uncontrolled deficits?

Which health plan should we choose for our company?

The answers to such questions are most often given in analytical reports. When the report is simple and short, it is usually presented in a memorandum or letter. (These types of reports are discussed in Chapter 11, Formats of Correspondence.) When the report is complex and long, it needs more structure, such as a title page, table of contents, and summary. These additional elements are not added to increase the weight and formality of the report; rather, they are added to help readers find their way through the report. (Chapter 9, Elements of Reports, provides the information you need to construct the elements of long reports. Here, the discussion is about how to put the elements into an appropriate format.)

Format

Depending on your purpose, the needs of your readers, and the content itself, analytical reports will use formats such as the following:

Format I	Format II
Title page	Title page
Table of contents	Table of contents
Executive summary	Introduction
Introduction	Summary
Discussion	Conclusions
Conclusions	Recommendations
Recommendations	Discussion

In Format I, although the executive summary highlights major conclusions and recommendations, the emphasis falls on the discussion. In Format II, the discussion is deemphasized and the summary, conclusions and recommendations are brought to the forefront.

Both are good formats. The one you would choose would depend on your needs and those of your readers. For example, you might know that your readers will be skeptical of your conclusions, and, therefore, wish to emphasize your discussion by choosing Format I. Or you might know that your readers will prefer seeing the big picture first, and, therefore, choose Format II. To formats like I and II, you can add, as needed, such elements as lists of illustrations, glossaries, documentation, and appendixes (all explained in Chapter 9, Elements of Reports).

Discussion sections of analytical reports can be rather specialized, depending on the questions being examined. The following should help you sort things out.

Discussion Sections

Discussion sections in analytical reports tend to use one of the following organizations: *classical argument, pro and con, choice of alternatives,* or *problem/solution.* No matter which organization you choose, your discussion has to build a firm base for the conclusions, recommendations, and decisions that follow from it.

In a *classical argument* format, you support a large opinion by a series of smaller opinions, which are in turn supported by facts. In argument, the large opinion is called the *major premise;* the smaller opinions are called *minor premises.* Your discussion format might look like this:

Major premise

  Minor premise A

    Factual support

  Minor premise B

    Factual support

  Minor premise C

    Factual support

In an actual argument, your format might look like this:

Uncontrolled U.S. federal deficits will lead to a decline in the worth of the dollar.

- As a result of foreigners financing U.S. deficits, too many dollars are in worldwide circulation.

    Factual support

- Worldwide, central banks are replacing dollars with yen and marks in their reserve portfolios, which lowers the value of the dollar.

    Factual support

- Deficits make investment in U.S. stocks and bonds less attractive for foreign investors, further driving down the value of the dollar.

    Factual support

Other forms of argument are all variations of the classical argument format.

A *pro and con* argument weighs the points in favor of something (the *pros*) against the points opposed (*cons*), resulting in a simple format:

A statement or a question
   Pro: Opinions and facts supporting the affirmative
   Con: Opinions and facts supporting the negative

An actual pro and con argument might break out like this:

Will the weakening of the dollar harm the U.S. economy?

Pros

- Loss of foreign investment lowers the value of U.S. stocks and bonds, which in turn lowers the value of pension funds.

    Factual support

- Weakening the dollar raises the price of foreign goods in the United States, resulting in inflation.

    Factual support

Cons

- Weakening the dollar increases the sale of U.S. goods in foreign markets.

    Factual support

- The higher cost of foreign goods in the United States lowers the competition for U.S. manufacturers.

    Factual support

After weighing the pros and cons for your readers, you will be expected, of course, to arrive at a conclusion: Weakening the dollar will harm the U.S. economy, yes or no?

Frequently, an analytical discussion calls for a *choice of alternatives.* You might be called on to make a recommendation concerning some major company purchase—for example, vans to make company deliveries. In a choice-of-alternatives plan, you have to deal with the alternatives available and the criteria by which you judge the alternatives. *Criteria* are the standards you use to judge something. In the case of vans, you might have as alternatives all the vans made by major truck manufacturers. The criteria might concern initial cost, operating cost, carrying capacity, and maintenance record. You can organize your discussion by either alternatives or criteria:

By Alternatives	**By Criteria**

By Alternatives

Van A
  Initial cost
  Operating cost
  Carrying capacity
  Maintenance record
Van B
  Initial cost
  Operating cost
  Carrying capacity
  Maintenance record

By Criteria

Initial cost
  Van A
  Van B
Operating cost
  Van A
  Van B
Carrying capacity
  Van A
  Van B
Maintenance record
  Van A
  Van B

Organizing by alternatives has the advantage of offering a complete discussion of each van in one section. Organizing by criteria has the advantage of allowing for selective reading; that is, some readers may be more interested in cost than carrying capacity. Organizing by criteria allows them to find and read the section they are most concerned with. Both plans are good. As usual, your purpose and audience will help you choose the one suited to your situation.

In the first part of a *problem/solution* discussion, define the problem. Use the available data to demonstrate that a problem really exists. For example, suppose you are an executive with a computer company that has a problem with its technical help line. Customers who call it get repeated busy signals. And after connecting to the help line, they may have waits up to an hour. To define the problem, you answer questions like these:

Typically, how many times does a buyer of one of your computers call in seeking help?

How many calls a day does your help line receive?

How many technical support consultants does the company have?

What is the cost to your company of running the help line?

Can the company afford to increase the service to an acceptable level and keep its profit margin high enough to stay in business?

How is the problem damaging customer relations?

And so forth.

After you define the problem, offer your solution. If there are criteria you must apply to any solution, clearly state them. For example, in the technical help line problem, a criterion might be that the solution must be affordable to the company. A solution that cut too heavily into company profit margins—such as hiring large numbers of consultants—would not be acceptable. Perhaps in this case, your solution might be to allow calls only to customers who buy service policies with their computers. The money from the service policies would pay for an expanded and acceptable technical help line.

After stating your solution, you would need to demonstrate its likely effectiveness. Again, you would be using your data to answer questions:

Have other companies tried this approach?

How successful have they been?

How much money is needed to expand the help line to an acceptable level of service?

How much would a service policy have to cost?

How would consumers react to buying such a policy?

and so forth

Your discussion has to show the strong likelihood of your solution being successful.

If you offer more than one solution, the solution portion of the report might use a choice-of-alternatives plan. In this case, you might offer two alternatives: a service policy and pay-as-you-go. In the pay-as-you-go plan, the customer would pay for each help call made. You might then compare the alternatives using criteria such as customer acceptance, ease of administration, and effect on profit margin. At the end of your discussion, you could weigh the evidence in a series of conclusions and recommend which solution the company should choose.

Using the various forms of argument to analyze a set of facts, you can carefully build a powerful case for your position. But remember that, ultimately, your analysis can be no better than your facts. Also remember that, even though you may be trying to persuade your readers to accept your point of view, it is your responsibility to argue ethically. Do not, for example, slant your facts one way or the other.

(See Appendix A for a sample analytical report.)

Proposals

In a proposal, one organization (or sometimes an individual) offers, for a price, its services to another organization. For example, Organization X, a research organization, may offer to research and offer a solution to a problem that Organization Y has. Or Company A, a computer software manufacturer, may offer to research the software needs of Company B and offer to install the software needed to satisfy those needs. A *proposal* is essentially a specialized form of argument and uses many of the techniques discussed on pages 114–18.

Proposals are either *solicited* or *unsolicited*. In a solicited proposal, an organization in need of services advertises its needs in a document called a *request for proposal* (*RFP*). The RFP will state the needs and request that organizations that can satisfy those needs submit proposals. The RFP will usually state quite specifically the format that such proposals must take, sometimes right down to the headings to be used. When such is the case, follow the instructions given in the RFP point by point. Not to do so will likely result in a rejected proposal.

In an unsolicited proposal, an organization sees a problem that another organization has and offers to provide the solution through its services. The following section describes a format that could be used in an unsolicited proposal, and the section following that describes a format that a student could use to propose a project to a teacher.

Unsolicited Proposals

Short proposals may look like correspondence. Longer proposals may need title pages, tables of contents, and so forth (all described in Chapter 9, Elements of Reports). In either case, the central format of the proposal will look something like this:

Project summary
Project description
   Introduction
   Rationale and significance
   Plan of work
   Facilities and equipment
Personnel
Budget
Appendixes

A *project summary* is essentially an executive summary (see pp. 90–94). In it, you briefly summarize your proposed services and emphasize the objectives of your proposal. Be sure to highlight how meeting those objectives will satisfy the needs of the client organization.

A *project description* comprises six sections: (1) introduction, (2) rationale and significance, (3) plan of work, (4) facilities and equipment, (5) personnel, and (6) budget.

1. Be sure your *introduction* (see pp. 94–97) makes clear the services you are proposing and how the successful outcome of your proposal will benefit your proposed client.

2. In *Rationale and Significance,* define the problem and make clear the need for a solution, describe the solution, show that the solution is feasible, and give the benefits of the solution. This section has the characteristics of an analytical report (see pp. 113–18).

3. To carry out your solution, you must have a *plan of work.* A plan of work section comprises smaller elements that state your scope, describe your methodology, break your work into its component tasks, and schedule your work:

- Describe the scope of the work to be done; that is, make clear what you will do and, sometimes, what you will not do. Being careful about describing scope may prevent later difficulties with clients who expect work that they think you have promised.
- Describe your methodology. For example, will you research further the problem you perceive in the organization? If so, what methods will you use? Will you, perhaps, use focus groups and questionnaires? If so, how will you evaluate the results you get through these methods? Show your potential clients you know what you're doing.
- The proposed work can no doubt be broken down into smaller tasks. For example, you may first give a test questionnaire to a small group. After evaluating the results, you may modify the questionnaire and give it to all the client's employees. Next, you may evaluate the results, and so on.

When you have made your work plan and its component tasks clear, give your schedule for completion of that plan. You may find a flowchart (see pp. 53–55), showing your tasks in relation to the time it will take to accomplish them, useful for clarifying your schedule.

4. In the *facilities and equipment* section, tell your prospective clients what facilities and equipment you will need and how you plan to get access to them. Answer questions like these: If you need a certain kind of laboratory, do you already have use of it? Do you have the equipment needed? If not, how do you plan to get it? Who will pay for it, you or the client? If the client will pay, will you use this equipment exclusively for the client? If not, how big a share must the client pay? You have a legal as well as an ethical responsibility to state clearly the answers to such questions.

5. In the *personnel* section, list the people who will work on the project, if the proposal is accepted. Give details of their relevant education and experience, such as the dates of past projects of a similar nature, the names and addresses of previous clients, and publications in the field of the proposal. The more detailed you can be about relevant qualifications, the better the chance of your proposal being accepted.

6. Finally, you present your *budget.* In a table or list of some sort, itemize your costs. If you have an extensive budget, you may need a classification scheme—for example, equipment, laboratory costs, salaries, travel, fee, and so forth.

In addition to these six sections, you may need *appendixes* to your project description (see pp. 99–100). Appendixes may include additional budget information, biographical information, company background, histories of earlier successful projects, and so forth. In making your selection, remember that your proposal is a sales document that is being read by busy people. They will read relevant appendixes but be put off by anything that looks too much like boilerplate.

Student Proposals

Students frequently must propose projects of various sorts to their teachers. For instance, a student in an advanced biology class may have to propose an experiment that runs over the semester. A student in a writing class may have to propose a project such as a feasibility report. If you are a student and a teacher gives you a plan for the proposal, follow it carefully. In the absence of such instructions, a scaled-down version of the unsolicited proposal will be appropriate. The following is an accepted format:

- Combined executive summary (see pp. 90–94) and introduction (see pp. 94–97) that makes clear the subject, purpose, scope, and benefits of the project
- Task and time schedule
- Resources needed and where they are available
- Your qualifications for carrying out the project

(See Appendix B for a sample student proposal.)

Progress Reports

If you are working for a client, he has a natural interest in the answer to the question: How are you doing? Progress reports are designed to answer that question. Some work, particularly work that results from an accepted proposal, requires progress reports at stated intervals, perhaps monthly. Whether written to schedule or at irregular intervals, progress reports follow fairly standard formats.

What follows is one such format. Like all reports, if it is only a few pages long, write the progress report as a memo or letter. If it is long and involved, add title pages and the like, as needed.

Introduction
Use a standard introduction (see pp. 94–97). Make clear what work you are reporting.

Project Description
In a project description, briefly describe the work being done, being sure to clearly state its purpose and scope. The scope statement breaks the work down into its component tasks—for example, devise questionnaires, administer questionnaires, evaluate questionnaire results, write final report.

Work Completed

Tell the reader what you have accomplished to date. In a long-running project, requiring several progress reports covering several periods, you might divide this section further as follows:

> Summary of work accomplished in preceding periods
> Work accomplished in reporting period

You may further subdivide these sections by the tasks you have indicated in your scope statement, like this:

> Work accomplished in reporting period:
>> Devising questionnaires
>> Administering questionnaires
>> Evaluating questionnaires
>> Writing final report

Work Planned for Future Periods

> Work planned for next period
> Work planned for future periods

Appraisal of Progress

Evaluate your progress. Indicate where you are ahead of plan and where you are behind. Don't offer a litany of excuses, but if there are good reasons as to why the work is not going to plan, state them clearly. The executive summary (see pp. 90–94) is a good model for this appraisal.

As in all writing, don't complicate your progress reports any more than necessary, but do answer thoroughly a client's three basic questions:

1. What have you done?
2. What are you going to do next?
3. How are you doing?

(See Appendix C for a sample progress report.)

Empirical Research Reports

If you pursue a scientific career, you will frequently have to report the results of your empirical research. In Chapter 8, Purposes of Formats, the empirical research report is used as an example. Here, you will find the typical format of such reports. You should also study carefully the reports published in the journals that cover your field of study. (See also Appendix D for a complete sample report.)

Abstract
The abstract summarizes key points of the report, including objectives, major results, and conclusions of the research.

Introduction
The introduction describes the subject, scope, significance, and objectives of the research.

Literature Review
The literature review summarizes previous research that has bearing on the research being reported. Such research may bear on the objectives, materials and methods, and rationale for the work done. In journal articles, the literature review is most often integrated with the introduction.

Materials and Methods
The materials and methods section may describe any of the following: the design of the research, materials used, procedures followed, and methods for observation and evaluation. An experienced researcher in the field should be able to replicate the research using this section as a guide.

Results
The results section is a factual accounting of what the researcher has found. Writers of research reports make extensive use of tables and graphs in the results section.

Discussion

The discussion section is an analytical discussion that interprets the results (see pp. 113–18). It answers questions such as these: Were the research objectives met? If not, why? How well do the results correlate with results from previous research? If they do not correlate well, why? What future work should follow this research? In many formats, the results and discussion are combined.

Conclusions

If the researcher's conclusions are not integrated into the discussion, they are stated here.

ENDNOTE

[1]David L. Wenny, "Calculating Filled and Empty Cells Based on Number of Seeds per Cell: A Microcomputer Application," *Tree Planters' Notes* 44 (1993): 49.

11

Formats of Correspondence

Letters are used for correspondence outside an organization. *Memorandums* (or *memos*) are used for correspondence within an organization.

Letters and memos may (and in some cases should) contain many of the elements found in more formal reports (as described in Chapter 10, Formats of Reports). That is, they may contain introductions, summaries (particularly executive summaries), discussions, conclusions, and recommendations. However, these elements may be somewhat abbreviated. They also may be labeled differently—"Findings" or "Discussion," for instance—or have substantive headings, such as "Existing Patient Populations." Such labeling is particularly appropriate for letters and memos that substitute for longer, more formal reports. (See Appendix A for an example of a letter report.)

Shorter letters and memos that convey information or opinions, make complaints, answer complaints, and so forth may have simple introductions and perhaps even summaries but will not label them as such.

The first section of this chapter deals with the formal elements used in standard letters. The second, third, and fourth sections deal with some important stylistic elements in simplified letters, memos, and e-mail. The closing section describes the correspondence used in a job hunt: résumés and letters of application.

126

Formal Elements of Letters

One of the disadvantages of having a personal computer on every desk is that executives in middle-management ranks are sometimes responsible for producing their own correspondence. Therefore, the formal elements of correspondence may be more important to you than you would like to believe. Also, good page design (see Chapter 5) is as important in letters as in reports.

Both Appendix A and Figure 11.1 illustrate well-formatted business letters. The following explains the components found in such letters.

Letterhead

On the job, you will likely have a printed letterhead on your stationery, containing your organization's address, telephone and fax numbers, and the like. Figure 11.1 shows a typical printed letterhead. If you don't have a printed letterhead, as might be the case when you are job hunting, simply place your address flush left, as illustrated in Figure 11.7 (p. 141).

Notice in the figures that standard postal abbreviations are used for states and provinces (see Figure 11.2). However, words such as *Street, Road,* and so forth are not abbreviated.

Date Line

Use one of these two styles for date lines: *16 March 1996* or *March 16, 1996.* Notice that the month is not abbreviated and number suffixes, such as *th* and *nd,* are not used.

Inside Address

Place the name and address of the person receiving the letter four spaces below the date line. Use a courtesy title, such as *Ms.* and *Professor,* with the name. The abbreviations *Mr., Ms.,* and *Dr.* are standard usage. Other courtesy titles, such as *Professor* and *Captain,* are spelled out. When you use the courtesy title *Dr.* before the name, do not use the equivalent (for example, *Ph.D.*) after the name. Put a one-word job title, such as *Director,* after the name. Place a job title of two or more words on the next line after the name.

B R O W E R   C O N S T R U C T I O N   C O M P A N Y

1998 Lee Highway
Freedom, MO 63032
Phone: 314-555-6788
Fax: 314-555-3097

July 27, 1996

Mrs. Irma Weaver
2 Brightside Lane
Freedom, MO 63032

Re: Your letter of July 17, 1996

Subject: Proposed solution of underground tank problem

Dear Mrs. Weaver:

We regret the installation of the underground oil tank on your
property when we built your house. We did, indeed, receive
the notice from the town clerk about the new city ordinance
prohibiting underground oil tanks for environmental reasons.
Unfortunately, for some reason, the news about the ordinance
did not reach our design team in time.

However, the town clerk tells us that during the winter everyone
with a building permit received a notice of the new ordinance.
Your permit number was 615002. Apparently, you overlooked the
notice, as did we.

FIGURE 11.1   Standard Business Letter

Page 2
Mrs. Irma Weaver
27 July 1996

I think we have a shared responsibility in this matter, and neither of us should bear the full cost. Let me suggest the following: We will remove the underground tank and fittings, close off and seal the opening in your foundation, and fill the hole at no labor or material cost to you. We will charge you labor costs for installing an indoor tank and the difference between the price of the outside tank and the new indoor tank. Your total cost will be $420.34.

If this arrangement is satisfactory to you, please call and we'll schedule the work. We regret any inconvenience you and Mr. Weaver have experienced and hope you will continue to enjoy your new house.

Sincerely yours,

Howard Brower

Howard Brower
Chief Operating Officer

HB: pgc

Encl. Copy of town clerk's letter to building permit holders

FIGURE 11.1 Continued

United States				Canada	
Alabama	AL	Missouri	MO	Alberta	AB
Alaska	AK	Montana	MT	British Columbia	BC
American		Nebraska	NE	Labrador	LB
Samoa	AS	Nevada	NV	Manitoba	MB
Arizona	AZ	New Hampshire	NH	New Brunswick	NB
Arkansas	AR	New Jersey	NJ	Newfoundland	NF
California	CA	New Mexico	NM	Nova Scotia	NS
Colorado	CO	New York	NY	Northwest	
Connecticut	CT	North Carolina	NC	Territories	NT
Delaware	DE	North Dakota	ND	Ontario	ON
District of		Ohio	OH	Prince Edward	
Columbia	DC	Oklahoma	OK	Island	PE
Florida	FL	Oregon	OR	Quebec (Province	
Georgia	GA	Pennsylvania	PA	de Quebec)	PQ
Guam	GU	Puerto Rico	PR	Saskatchewan	SK
Hawaii	HI	Rhode Island	RI	Yukon Territory	YT
Idaho	ID	South Carolina	SC		
Illinois	IL	South Dakota	SD		
Indiana	IN	Tennessee	TN		
Iowa	IA	Texas	TX		
Kansas	KS	Utah	UT		
Kentucky	KY	Vermont	VT		
Louisiana	LA	Virginia	VA		
Maine	ME	Virgin Islands	VI		
Maryland	MD	Washington	WA		
Massachusetts	MA	West Virginia	WV		
Michigan	MI	Wisconsin	WI		
Minnesota	MN	Wyoming	WY		
Mississippi	MS				

FIGURE 11.2   State, Territory, and Province Abbreviations for the United States and Canada

As in the letterhead, abbreviate names of states and provinces but not words like *Street.* Write the names of organizations and people exactly as they do. Use *Inc.,* not *Incorporated,* if the company does. Use *F. Xavier Jones,* not *Frank X. Jones,* if he does.

Re *Line*

The *Re* in the *re* line stands for "reference." Generally, the reference is to other documents, as in *Your letter of 12 April 1996* or *Your contract with Smith Brothers, dated 16 May 1995.*

Subject Line

In a subject line, you tell the reader what subject will be dealt with in your letter—for example, *Summer Schedule for Executive Committee Meetings.* Sometimes, the word *Subject* is used in the subject line, as in Figure 11.1. When *Subject* is omitted, the subject line is generally all capital letters, as in Figure 11.3.

Salutation

For the most part, we still adhere to tradition and begin salutations with *Dear.* Follow *Dear* with the name used in the inside address—*Dear Ms. Pleasant* and so forth. Use a colon after the salutation, and place it as shown in Figure 11.1.

If you do not have a name to use, you have one of two choices: If it's an important letter—a proposal, for instance—get a name, even if it takes a long-distance phone call to do so. If it's a routine letter—such as an inquiry—use a simplified letter, as shown in Figure 11.3.

Body

Keep body paragraphs short, rarely more than six or seven lines, and space them as indicated in Figures 11.1 and 11.2. Do not split words between lines. Also do not split dates or names; that is, *February 11, 1997* should be on one line, as should be *Margaret M. Briand.*

Complimentary Close

In most business correspondence, use a simple complimentary close, such as *Sincerely yours.* In instances in which friendships are involved, closes such as *Best regards* are suitable. Follow the close with a comma, and place it as in Figure 11.1.

B R O W E R   C O N S T R U C T I O N   C O M P A N Y
1998 Lee Highway
Freedom, MO 63032
Phone: 314-555-6788
Fax: 314-555-3097

26 March 1996

Director, Corporate Research and Engineering
Burnham, Inc.
3660 Folwell Drive
Minneapolis, MN 55418

SPREAD-SPECTRUM HOME SECURITY SYSTEMS

Our company installs a good many Burnham Home Security Systems
in the new homes we construct. These systems are hard wired,
which presents no problems in new construction. However,
increasingly, we are asked to provide security systems in existing
homes. Here, hard wiring presents problems for which the only
solution is running wires through walls, with all the accompanying
expense.

I understand that Burnham is working on wireless security systems
using a spread-spectrum modulation technique that will allow radio-
frequency communication between the components of a home
security system.

How far along is your research on this new system? Do you have a
target date for marketing it? When it is ready, we will certainly
consider it for use in both new construction and existing homes.

Howard Brower

Howard Brower
Chief Operating Officer

HB: pgc

FIGURE 11.3   Simplified Letter

Signature Block

Four spaces below the complimentary close, type your name and, if you have one, your title. To avoid complicating the life of a correspondent who doesn't know you, use enough of your name to indicate your gender. That is, use *Patrick M. Fields,* not *Pat Fields* or *P. M. Fields.* In the space between the complimentary close and your typed name, sign your name—legibly, please.

End Notations

Notations following the signature block indicate such things as identification, enclosures, and copies (see Figure 11.1).

Identification

In this type of notation, the writer's initials are in capital letters and the secretary's initials are in lowercase:

FDR/hrc

Enclosures

Enclosure lines indicate to the reader—sometimes in a general way, sometimes specifically—that you have enclosed additional material with your letter, as in the following two examples:

Enclosures (3)

Encl: Schedule for summer meetings

Copy

In a copy line, you tell your correspondent who is receiving a copy of the letter:

cc. Dr. Georgia Brown

Mr. Hugh Binns

Continuation Page

When your letter exceeds one page, you need a continuation page or pages (see Figure 11.1). Follow these rules when constructing a continuation page:

- Use paper of the same color and weight as your first page, but do not use letterhead stationery.
- As in Figure 11.1, head the continuation page with the page number, the name of your correspondent, and the date.
- Have at least three lines of text on the last continuation page before the complimentary close and whatever else follows.
- The last paragraph on the page that precedes the last continuation page should contain at least two lines.

Simplified Letters

The simplified letter (see Figure 11.3), as its name implies, is a simplified form of the standard letter. It always has a subject line, but it does not have either a salutation or a complimentary close. In every other respect, it follows the format of a standard letter. In part, the simplified letter is a response to the fact that salutations such as *Dear Sir, Gentlemen,* and *To Whom It May Concern* are no longer in fashion.

Use a simplified letter only for routine correspondence to an organization in the case in which you do not have the name of a person to address. You could use it, for example, to register a complaint with an organization or to make a simple inquiry to some department within an organization. Do not use a simplified letter to answer complaints (where good strategy calls for you to address the person complaining by name) or for important letters like letter reports.

Memorandums

Memorandums are used for correspondence within an organization. They most often are written on printed forms that are headed with the organization's name and spaces for *date, to, from,* and *subject* (see Figure 11.4). Because of the memo's format, a salutation and signature block are not needed. The body of a memo and its continuation page look precisely like the body and continuation page of a letter. A memo also uses the same end notations as a letter.

Memos may be used for any of the purposes for which letters are used. That is, you can write memos that provide information or make inquiries or memo reports that, like letter reports, are short reports containing summaries, introductions, headings, and so forth.

Freedom Community Hospital

Date: 12 April 1996

To: Barbara Smith
           Purchasing Officer

From: John Gamez, M.D. *JG*
           Chief of Cardiac Surgery

Subject: Vascular Hemostasis Device

The FDA has approved a vascular hemostasis device for stopping bleeding after angioplasty and angiography. It seems promising. Please order several dozen for our division so that we can try them.

The device is made by Medico Corporation of Savannah, Georgia.

JG: akp

FIGURE 11.4   **Memorandum**

E-mail

The use of *e-mail* as a substitute for letters, memos, and phone calls is growing rapidly. Its speed and simplicity make it very attractive, both within and without organizations. E-mail format is determined by the online organization used but usually includes entries for address, subject, message, and so forth, similar to a memo.

The messages sent by e-mail tend to be brief. When used among trusted colleagues, e-mail is likely to be highly informal and full of abbreviations and other shortcuts, perhaps known only to those sending and receiving it. If you do not save the e-mail messages you send and receive, you may face the same pitfall as with phone conversations: That is, people's memories of what was actually said or implied in e-mail may differ widely.

E-mail is probably best used for sending needed information and opinions but not for the equivalent of letter or memo reports.

Correspondence of the Job Hunt

Begin your job-hunt correspondence by brainstorming the elements of your past education and job experience. With the help of college transcripts, memory, and any journals and records you had the foresight to keep, list details such as dates of employment, job titles, course titles, professors' and employers' names, work accomplished, and so forth. Make your list as extensive as possible. The information you list will furnish material for résumés and letters of application, the two most important pieces of job-hunt correspondence.

Résumés

A *résumé* is a summary of your education and job experience that you send to potential employers. From it and the accompanying letter of application, potential employers will decide whether to interview you. Thus, your résumé is a very important piece of paper. See Figures 11.5 and 11.6 for example résumés.

Word process your résumé and print it on good-quality white paper, using a letter-quality printer. Take advantage of the font possibilities on a word processor, but use discretion and good page design (see Chapter 5). Don't fill your résumé with exotic typefaces. The mix of plain print and boldface in Figures 11.5 and 11.6 illustrates the look you want. Résumés must be mechanically perfect—no misspellings, typos, or grammatical errors. If you need help to accomplish that, get help.

Figure 11.5 illustrates a chronological résumé, and Figure 11.6, a functional résumé.

Chronological Résumés

Look at Figure 11.5. Begin a chronological résumé with your name and list all the ways you can be reached: by mail, phone, fax, and e-mail. Next, give the details of your education: degree, expected graduation date, major, minor, course work, extracurricular activities, and so forth. Avoid the constant use of *I* by using fragmentary sentences.

RÉSUMÉ OF JANICE OSBORN

32 Merchant Road
St. Paul, MN 55101
Phone: (612) 555-6755
E-mail: josborn@aol.com

Education
1992–1996   **University of Minnesota, St. Paul, MN**
Candidate for bachelor of science degree in technical
communication with a minor in computer science in
June 1996. In upper-third of class with a GPA of 3.0 on
a 4.0 scale. Member of St. Paul Student Council; vice
president in senior year. Served as editorial assistant,
1994–1996, for *Technical Communication Quarterly,* the
journal of the Association of Teachers of Technical Writing. Corresponded with authors and copyedited articles.

Business
Experience
1995   **Communication Design Associates,**
Summer   **Minneapolis, MN**
As part of five-person team, assisted large corporation in
using SGML (Standard Generalized Markup Language)
to convert 10,000 pages of paper documentation to
online documentation.

1993–1994   **Technical Publications, Inc.**
Summers   Using desktop publishing techniques, worked collabora-
tively and individually in developing manuals, proposals,
and feasibility reports.

Personal   Grew up in International Falls, Minnesota.
Background   Have traveled in United States and France.
Enjoy music, Alpine and Nordic skiing, and browsing
Internet.

References   References available on request.

March 1996

FIGURE 11.5   Chronological Résumé

You should indicate your academic standing in the most favorable way you legitimately can. If your GPA in your major is higher than your overall GPA, use that. If your record is really bad, don't list it. If you attended more than one college, list the last one first and so on.

Give the details of your work experience. As with your education, do it in reverse chronological order. Don't merely list job titles. Using action verbs like *managed, operated, organized, sold,* and *designed,* describe what you did. List all the jobs of your college years, even those that don't relate to the jobs you're seeking. Employers feel, quite correctly, that people who have worked understand the workplace better than those who have not.

If you have room, give a few details of your personal background. Sometimes, employers will see something there that interests them, such as people skills, which they value highly.

Offer to supply references. Use both professors and employers (more on this later). Finish with the month and year of the résumé.

The advantage of the chronological résumé is that it provides a smooth summary, year by year, of your education and experience. If growth is there, this type of résumé will show it well. But if your education and experience have big gaps, the chronological résumé may not be the best choice. Also, your skills and aptitudes may tend to get lost in the welter of dates, courses, and jobs. Despite all that, the chronological résumé is a good format.

Functional Résumés

Look at Figure 11.6. Begin a functional résumé with your name, address, phone number, and the like, as in the chronological résumé. Next, give your degree, major and minor, and expected graduation date. If you attended more than one college, give them in reverse chronological order.

The heart of a functional résumé is a classification of the experiences—academic, extracurricular, and work—that demonstrates your skills and capabilities. (See pp. 16–17 for the rules of classification.) Using category words such as *professional, technical, people, communication, management, marketing, sales,* and *research,* create two or three categories that show you off the best.

Finish the résumé with a reverse chronological listing of your jobs, an offer of references, and the date.

RÉSUMÉ OF JANICE OSBORN

32 Merchant Road
St. Paul, MN 55101
Phone: (612) 555-6755
E-mail: josborn@aol.com

Education   Candidate for bachelor of science degree in technical communication with a minor in computer science from University of Minnesota in June 1996.

Professional
- Working under pressure, used desktop publishing to produce high-quality manuals, proposals, and feasibility reports.
- As part of five-person team, assisted large corporation in using SGML (Standard Generalized Markup Language) to convert 10,000 pages of paper documentation to online documentation.
- Assisted editor of professional journal in copyediting manuscripts and corresponding with authors.
- Completed courses in writing, editing, speaking, desktop publishing, graphics, management, multimedia, and computer science.

People
- Elected vice president of St. Paul Campus Student Council. Oversaw student recreational budget.
- Successfully worked in collaboration with other writers and editors.
- Know how to accept criticism and use it constructively.

Work Experience

Summer, 1995
- Communication Design Associates, Minneapolis, MN: Writer and member of consulting team.

1994–1996
- *Technical Communication Quarterly*, St. Paul Campus: Editorial Assistant.

1993, 1994 Summers
- Technical Publications, Minneapolis, MN: Technical writer and editor.

References   References available on request.

March 1996

FIGURE 11.6   Functional Résumé

The advantage of the functional résumé is that it brings to the fore your skills and capabilities. There is a slight disadvantage in that it does not show the smooth progression of the chronological résumé. Choose the format that displays you to your best advantage.

Letters of Application

When you send out your résumé, accompany it with a letter of application (see Figure 11.7). A letter of application is a letter of transmittal for the résumé, but it is also a place where you can highlight your capabilities and catch an employer's interest. If you are discreet about it, you can use your letter of application to point out how you could fit into the organization and why it would be to their advantage to hire you. Blow your own horn but not directly in the employer's ear.

If possible, send your letter and résumé to the person for whom you might work. A potential supervisor will be better able to judge your qualifications than a human resources officer. Often, a letter or phone call to the organization might get you the name you need. Looking through professional journals in your field will often provide such names, as will networking with professionals. Websites and forums on the Internet also can provide useful information about employers. Send letters and résumés to human resources divisions if need be but only as a last resort.

A good way to start your letter of application is by dropping a name known to the potential employer but only with permission. Indicate some knowledge of the organization, which you can gather in the same way as names. In the middle of the letter, *sell yourself.* Use specific education and work experiences to point out your potential value and usefulness to the organization.

In the close of the letter, refer to your résumé and references. If you have significant products from your work or education (such as research reports, manuals, videos, and so forth), offer to send them. Finally, try to arrange an interview. Make it as convenient for the employer as you possibly can.

Other Correspondence

Several other pieces of correspondence are necessary during the job search, none of them difficult or time consuming to do.

32 Merchant Road
St. Paul, MN 55101

9 March 1996

Mr. James Cantrell
Supervisor of Writing and Publications
Bell Computer Corporation
4200 Lake Avenue
Madison, WI 53714

Dear Mr. Cantrell:

Professor Robert Wilson of the University of Minnesota Computer Science Department tells me that your firm designs communication protocols that allow two or more computer networks to operate as one network. This is a field of great interest to me, and I think I have the education and experience to serve Bell Computer well.

In June, I will graduate from the university's technical communication program with a minor in computer science. I have completed courses in writing, editing, speaking, desktop publishing, graphics, management, multimedia, and computer science.

In the summers of 1993 and 1994, I worked as a technical writer and editor using desktop publishing to produce high-quality manuals, proposals, and feasibility reports. Last summer, I worked on a team helping a large corporation use SGML to convert paper files into online files. This experience, combined with my education, would allow me to fit into your operation quickly.

The enclosed résumé gives more detailed information about my education and experience. I can provide references from both my teachers and employers.

May I drive over to Madison to discuss job opportunities with you? If that is not convenient, I'll be attending the International Technical Communication Conference in May. Perhaps we could talk there.

Sincerely yours,

Janice Osborn

Janice Osborn

FIGURE 11.7   Letter of Application

When you ask your references for permission to use their names, provide them with copies of your résumé. If you can, call on them. If you can't, write each person a letter, recalling your relationship with him or her and asking for permission.

Sending several thank-you notes is appropriate during and after the job hunt. When you interview, get the interviewer's name and address and write a note, expressing your appreciation for the interview. When your job search succeeds, write thank-you notes to your references. Give them the outcome of the search, and thank them for their help. They'll be curious about the result and pleased with your thoughtfulness.

If you are offered jobs at several organizations, drop notes of thanks and refusal to the organizations you turn down. Simply express your appreciation for the job offer, thank them for their time and interest, and perhaps compliment them on their organization.

APPENDIXES

Sample Reports

A   Letter Analytical Report

B   Student Proposal

C   Progress Report

D   Empirical Research Report

APPENDIX A

Letter Analytical Report

Analytical reports up to six or seven pages long are likely to be written as correspondence, as is the report in this sample. The sample uses a problem-solution organization. In the interests of accessibility, selective reading, and comprehension, it contains a good introduction that makes clear the subject, purpose, and scope of the report. Headings also aid selective reading and display the report's organization. The Recommendation section makes good use of listing techniques.

Environmental Consultants

2063 Peach Tree Street
Suite 260
Atlanta, GA 30747
Phone: 404-555-1940
Fax: 404-555-9003
E-mail: EnCo@aol.com

November 20, 1996

Mr. James Morris
Chief Executive Officer
Albany Office Products
22 Oglethorpe Road
Albany, GA 30278

Subject: Inspection Findings and Recommendations

Dear Mr. Morris:

As you engaged us to do, we have examined the environmental
problems in your company headquarters. This report provides
background for our report, states the problems our inspection
uncovered, and recommends solutions for them.

Background

For the last year, the workers in your two-story office building
have experienced higher-than-average health problems. They
have suffered from watery eyes, nasal congestion, coughing,
difficulty breathing, headache, and fatigue. Last winter, your

Page 2
Mr. James Morris
November 20, 1996

workers had a high incidence of flu, and lost work days grew to unacceptable levels because of it. During the last six months, three of your employees with asthma had to be hospitalized.

Although these problems were not limited to your first-floor offices, they were more prevalent there than on the second floor. All these reactions pointed to the presence of excessive biological pollution in your building, particularly on the first floor.

Biological pollutants are found everywhere. Molds, bacteria, and viruses are commonly found in office buildings such as yours. People exposed to such pollutants may suffer allergic reactions, infections, and even serious toxic reactions in the central nervous system and the immune system.

Biological pollutants need moisture to grow and spread. When moisture levels in a building are lowered, the level of pollutants and the reactions to them are greatly reduced. Therefore, our inspection of your building focused on moisture problems, particularly on the first floor.

Findings of the Inspection

Our inspection found major problems with your first-floor carpet and heating and air-conditioning ducts. We found also a minor problem with the large number of coffee makers in your building.

First-Floor Carpeting

Your building is slab constructed, and bare concrete underlies all of the matting and wall-to-wall carpet on the first floor. Moisture has passed through the concrete and allowed mold to grow and spread in and under the matting and the carpet. Spot inspections indicate that more than 80 percent of the first-floor carpet has mold growing underneath it. Where mold is found, you can be sure that bacteria and viruses also flourish. This condition most certainly explains the high incidence of health problems on the first floor.

Heating and Air-Conditioning Ducts

The heating and air-conditioning ducts are full of dust and are beginning to show signs of mold. The heater and air conditioner had not been properly cleaned, and the system filters were clogged with dust. These conditions explain the incidence of health problems on the second floor.

Coffee Makers

Office policy obviously does not regulate the use of coffee makers throughout the building. We found 18 coffee makers plugged in and working during our inspection. These coffee makers put a great deal of moisture into the air, encouraging the growth of pollutants.

Recommended Solutions

The problems encountered in your building all have ready solutions.

First-Floor Carpeting

The first-floor carpeting and matting are too far gone to be salvaged. They must be taken up and discarded. The concrete floor must be professionally cleaned and disinfected. Following that, you have two alternatives:

Page 4
Mr. James Morris
November 20, 1996

- Lay a plastic vapor barrier on the concrete and cover that with a subfloor of insulation and plywood. Matting and wall-to-wall carpet can then be laid on the plywood.

- A somewhat less expensive alternative would be to lay good-quality asphalt or vinyl tile on the concrete and use area rugs where carpeting is wanted.

Heating and Air-Conditioning Ducts

You need to take three steps to keep your heating and air-conditioning ducts free of pollution:

- Have the heating and air-conditioning units and their ducts professionally cleaned as soon as possible but not until after the carpet problem has been resolved. (Removing the carpet, cleaning the concrete floor, and laying subflooring or tile will kick up dust and dirt.)

- Contract with heating and air-conditioning professionals to have them clean your heaters and air conditioners at the start of each heating and cooling period.

- Arrange to have your building cleaners change your system filters monthly.

Coffee Makers

Remove the coffee makers from the various offices. If you wish, in some well-ventilated space, place one or two larger coffee makers that everyone can use.

Page 5
Mr. James Morris
November 20, 1996

The recommended solutions follow the guidelines laid down by the U.S. Consumer Product Safety Commission and the American Lung Association. While air pollution cannot be completely eliminated, short of extraordinary measures, carrying out the work recommended will restore a healthful environment to your workplace.

If you want our assistance in locating the professionals to carry out the needed work, let us know. Thank you for letting us help you.

Sincerely,

Nancy Larsen

Nancy Larsen
Chief of Inspections

NL: siu

Student Proposal

Although a *student* proposal, this sample provides much the same information that might be found in a full-scale proposal. The introduction makes clear the purpose and scope of the work to be done, its rationale, and its significance both to the professor's proposed research and to the student. The Time and Task Breakdown section shows the student's plan of work. The Resources Available section parallels the more full-scale Facilities and Equipment of a large proposal. The listing of the journals and books available shows that the student has done a preliminary study, always a good sign to the person expected to approve a proposal. The Qualifications section, like a Personnel section, gives details of the student's education and experience that qualify her for the work proposed. Only the Budget section of a full-scale proposal is missing (and not needed) in this proposal.

MEMORANDUM

Date   18 August 1996

To   Professor Donald Hood, Chair
              Department of English
              University of Southern Idaho

From   Madison James *MJ*

Subject   Proposal for an Independent Study

Professor Elizabeth Irvin, head of the university writing laboratory, plans a mentoring research project during the winter term. In the project, university undergraduate English majors using the university's computer network will mentor freshman students in writing . The project's two major objectives will be to study the effect of the mentoring on the writing skills of the freshmen and to study the effect of the mentoring process on the mentors themselves. For example, will the act of mentoring deepen the mentor's understanding of the writing process?

I have volunteered to help Professor Irvin by reviewing literature pertinent to the project. The review will focus on the methodology and results of related research. It should help Professor Irvin plan her methodology and sharpen her objectives. Conducting the search and reporting its results should increase my own research and writing skills.

I request that I be allowed to use this literature search and the resulting literature review to fulfill the requirements for three hours of independent study in English 4500. Professor Irvin will guide me in the search and be the instructor of record for the independent study.

Page 2
Professor Hood
18 August 1996

Task and Time Breakdown
In order to be useful in Professor's Irvin's planning, the literature review should be in her hands by mid-November. Therefore, I propose the following timetable:

1. Search the periodicals and books in the university library for literature relevant to Professor Irvin's project. Take notes on useful information. (4 weeks)

2. Write and submit a progress report on my search to Professor Irvin, with a copy to you. (1 week)

3. Continue the library search, and begin preliminary organization and analysis of the information gathered. (2 weeks)

4. Write a preliminary draft of the literature review, and submit it to Professor Irvin, with a copy to you. Discuss the preliminary draft with Professor Irvin. (2 weeks)

5. Write the final draft of the literature review, and submit it to Professor Irvin, with a copy to you, no later than 15 November.

Resources Available
The university's periodical collection in the area of this search is comprehensive. The following periodicals seem particularly relevant: *American Educational Research Journal, College Composition and Communication, College English, Computers and the Humanities, J. of Educational Research, J. of Experimental Education, Research in the Teaching of English, Teaching English in the Two-Year College,* and *Writing Lab Newsletter.*

Page 3
Professor Hood
18 August 1996

The National Council of Teachers of English has published many books on writing and some on the use of computers in teaching writing. *Computers in English and the Language Arts* (Editor, T. Batson) and *Writing and Response: Theory, Practice, and Research* (C. M. Anson) seem especially useful. I'm sure further search will uncover many more sources.

Qualifications
As a senior student in an English major, I have conducted numerous library searches. I have been a part-time employee in the writing laboratory for three semesters and understand its methodology and objectives. I successfully completed a three-hour independent study last year with Professor John Englebart. My English GPA is 3.8.

cc: Professor Elizabeth Irvin

APPENDIX C

Progress Report

This progress report is the one called for by the proposal in Appendix B. Like all good progress reports, it answers the three questions every client wants answered: What have you done? What are you going to do next? How are you doing? The introduction and Project Description remind the client of what the subject and purpose of the research are. Under Work Completed, the writer provides enough detail to reassure the client that work is going well. The writer's dividing the subject matter into three areas suggests what the organization of the report will likely be. The Work Remaining section promises that the work will be done on time.

In the Overall Appraisal, the writer attempts something rather tricky and risky: She suggests, very tactfully, that a change in direction for some of the client's research might be in order. That, in itself, is not all that unusual, but for an undergraduate student to make such a suggestion is. It has to be done carefully, and it is.

MEMORANDUM

Date   1 October 1996

To   Professor Elizabeth Irvin, Head
University Writing Laboratory
University of Southern Idaho

From   Madison James   *MJ*

Subject   Progress Report on Independent Study for English 4500

I began work on this independent study on 1 September, after
you and Professor Hood accepted my proposal. This is the four-
week progress report scheduled in the proposal. It describes
the work completed and the work remaining in the study and
concludes with an appraisal of the progress of the study.

Project Description
The purpose of this independent study is to review the literature
pertinent to your mentoring research project scheduled for the
winter semester. The review focuses on the methodology and
results of the related research.

    This study has three major tasks:
- Search the periodicals and books in the university library for
  literature pertinent to the mentoring research project.
- Organize and analyze the information gathered.
- Write a literature review based on the information gathered.

Work Completed
I focused my search on three specific areas that match the
objectives and methodology of your research project: studies
that measured the effect of mentoring on the mentors, studies
that measured the effect of mentoring on the students, and
studies conducted via computer networks. I found excellent
articles in all three areas and one article that dealt with all three
areas. Here, I will briefly outline the results of the search in the

Page 2
Professor Irvin
1 October 1996

three areas. Because the article that deals with all three areas relates so closely to your project, I will discuss it at greater length.

Effect on the Mentors
All articles and books dealing with the effect on the mentors reported that the mentors (or *tutors,* the terms are used somewhat interchangeably) received significant benefits from their mentoring, increasing their skills in whichever fields they mentored in. Some of the more useful sources are listed here:

Allen, V. L., and R. S. Feldman. 1973. Learning through tutoring: Low achieving children as tutors. *J. of Experimental Education* 42: 1–5.

Vygotsky, L. S. 1987. *The collected work of L. S. Vygotsky.* New York: Plenum.

Witte, S., P. R. Meyer, T. P. Miller, and L. Faigley. 1981. *A national survey of college and university writing program directors* (Writing Program Assessment Project GRG 106-A). Austin, TX: University of Texas at Austin.

Effect on the Students
Although most studies report that mentoring has a beneficial effect on students, the results are less clear cut than those reporting the effect on the tutors. Useful sources:

Freedman, S. W., C. Greenleaf, and M. Sperling. 1987. *Response to student writing* (Research Report No. 23). Urbana, IL: NCTE.

Murray, D. 1979. The listening eye: Reflections on the writing conference. *College English* 41: 13–18.

Walker, C. P., and D. Elias. 1987. Writing conference talk: Factors associated with high- and low-rated writing conferences. *Research in the Teaching of English* 21: 266–285.

Computer Networks
There are now numerous reports of the use of computer networks to mentor students, including their use in writing laboratories:

Bump, J. 1990. Radical changes in class discussion using networked computers. *Computers and the Humanities* 24: 49–65.

Duin, A. H., E. Lammers, L. D. Mason, and M. F. Graves. 1994. Responding to ninth-grade students via telecommunications: College mentor strategies over time. *Research in the Teaching of English* 28: 117–153. (Because this study relates so closely to the one you propose, I will discuss it at greater length.)

Kinkaid, J. A. 1987. Computer conversations: E-mail and writing instruction. *College Composition and Communication* 38: 337–341.

Duin et al. Study
The Duin study closely parallels the study you propose. Its methodology section discusses training the mentors, student assignments, and data collection. Included in the data collected were the kinds and forms of mentor responses to students and information gathered from mentor journals and interviews with the mentors. Much of the methodology described, appropriately modified, should directly apply in your study.

The results show that the mentors worked hard at developing strategies for responding to the students and, in the end, developed well-structured responses that dealt with student needs. Mentors developed seven response categories, such as word choice, mechanics, organization, and idea development. The mentors reported increased confidence in approaching their own writing tasks as a result of their experience. The students' English teacher reported that her students gained both competence and enthusiasm during the study.

Work Remaining
The work remaining includes collecting more information, analyzing and organizing the information, and writing the literature review. I should have the first draft of the review ready by 21 October. Following discussion with you, I'll have the final report to you and Professor Hood by 15 November. I hope the review will be a well-documented summary of the major methodologies and results that will be useful in your study.

Overall Appraisal
The library search has gone well. I have copious notes that, when organized and summarized, should be of great use in your study. I have a suggestion about your study: One of the conclusions in the Duin et al. report is that there is a crucial need for a comparison study between the effects of computer mentoring and face-to-face mentoring.

Because the beneficial effects of mentoring on mentors has been so firmly established, it might be worthwhile to consider substituting a comparison study, such as Duin et al. suggests, for your objective of studying the effect of mentoring on the mentors. The writing laboratory would be an ideal setting for such a comparison study. If such a substitution seems reasonable to you, please let me know within a week, as it will affect both my organization and the first draft of the literature review.

cc. Professor Donald Hood

APPENDIX D

Empirical Research Report

Appendix D comes from a journal called *Tree Planters' Notes* (45 [1994]: 47–52). As you read it, pay attention to the following aspects:

- The completeness of the abstract
- The way in which the literature review leads into both the research methodology and objectives
- The clear statement of the objectives
- The organization and detail of the Materials and Methods section (A fellow expert could easily replicate this experiment.)
- The use of appropriate headings
- The predominant use of passive voice in Material and Methods and active voice in Results and Discussion
- The use of figures to summarize and clarify data
- The explanation and emphasis of key data in the discussion
- The way in which the conclusions meet the stated objectives
- The system of documentation used

Overwintering Black Spruce Container Stock Under a Styrofoam® SM Insulating Blanket

Robert E. Whaley and Lisa J. Buse

Stand establishment project forester, Northwest Region Science and Technology, Ministry of Natural Resources, Thunder Bay, Ontario, and education and communication coordinator, Ontario Forest Research Institute, Ministry of Natural Resources, Sault Ste. Marie, Ontario, Canada

In northwestern Ontario, large numbers of container seedlings are overwintered either outdoors or in frozen storage. Seedlings stored outdoors are subject to severe conditions overwinter and are prone to considerable damage, whereas freezer storage is expensive and has its own associated risks. To examine an alternative method for overwintering black spruce (Picea mariana (Mill.) B.S.P.) container stock, we tested different configurations of a rigid Styrofoam® SM insulating blanket. Third-year outplanting results showed no differences in growth or survival performance between controlled frozen and Styrofoam® SM stored stock. Tree Planters' Notes 45(2):47–52; 1994

Currently, about 159.2 million seedlings are produced for forest renewal in Ontario. Of these, about 55 million container and 44 million bareroot seedlings are overwintered outdoors. These seedlings can suffer considerable damage during outdoor overwinter storage, principally from root damage due to rapid freezing and shoot desiccation. Other problems include mechanical damage, such as the flattening of seedlings from the weight of snow and ice, and snow mold. All these problems are due to, or accentuated by, insufficient or fluctuating snow cover on outdoor stored stock and/or improper conditioning of the seedlings. Therefore, the emphasis in overwintering nursery stock is on protecting seedlings from desiccation and protecting seedling roots from critically low and/or rapid changes in temperature (McNiel and Duncan 1983).

Outdoor overwintered container stock suffered heavy losses in the Thunder Bay area in the years leading up to and including 1987 (OMNR 1987). These losses prompted the examination of alternative overwintering techniques, such as indoor frozen storage facilities, snow making, and insulating blankets. This report examines the feasibility of outdoor overwintering black spruce (*Picea mariana* (Mill.) B.S.P.) seedlings grown in Styroblocks under Styrofoam® SM insulating blankets.

2   *Tree Planters' Notes*

Insulating blankets have been in use throughout North America since the mid-1970's for protecting overwintered horticultural plants (Green and Fuchigami 1985). Their use in forestry for protecting container seedlings is a recent practice. Several different configurations of insulating blankets have been tried, such as placing polyethylene film over plants packed in straw or covering seedlings with a sandwich of straw between layers of either clear or translucent polyethylene film (Green and Fuchigami 1985). Other insulating materials used have included rigid Styrofoam® and manufactured "Microfoam" sheets that can be up to 19 mm thick (Gouin 1977).

The procedure for protecting container seedlings consists of either erecting a frame over seedling trays to support the blanket or aligning rigid trays on their sides and surrounding them with an insulating blanket, thereby sealing in the stored stock. The weight of winter snow and/or other material such as wooden strips holds the blanket down and keeps it from whipping in the wind.

To evaluate the use of a Styrofoam® SM insulating blanket for protecting overwintered container stock, we began tests at Hodwitz Enterprises Ltd. in Thunder Bay, Ontario, in 1990. The objectives of our tests were to:

1. Examine the feasibility of overwintering Styroblock seedlings under Styrofoam® SM insulating blankets.
2. Determine the cost of and procedures for overwintering stock using insulating blankets.

Materials and Methods

A demonstration project was established at Hodwitz Enterprises Ltd., Thunder Bay, Ontario, to investigate the ability of rigid Styrofoam® SM insulating blankets to protect seedlings from rapid and extreme changes in temperature, drying winds, and desiccating sun. The containers examined were Styroblock 130's and 165's, which measure about 47 by 35.1 by 13 cm (18 by 14 by 5 in) and hold 130 and 165 seedlings, respectively.

Test Configuration—Year 1. For the first overwintering period of this test (1990–91), 50 trays of Hodwitz Enterprises' black spruce Styroblock 165 seedlings were placed under an insulating blanket. This equated to approximately 8,200 seedlings.

The treatment trays were placed on their sides in pairs, with seedling tops together—the tray tops were about 25 cm (10 in) apart—and tray bottoms touching, in both single- and double-layer configurations (figure 1B). The "blanket" consisted of covering the trays with 5.1-cm-thick (2-in-thick) R10 Styrofoam® SM on the top and sides. The treatment seedlings remained covered from mid-November to mid-April. The control trays were stored as normal outdoor overwinter stock—uncovered and unprotected by cardboard boxes (figure 1A).

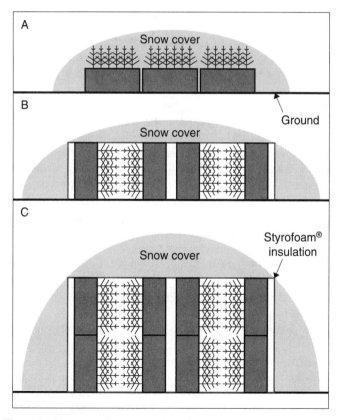

Figure 1—*Diagrammatic representation of overwinter storage treatments for black spruce in Styroblock 165's stored normally (**A**), one layer (**B**), and two layers deep (**C**) under the Styrofoam® SM blanket.*

Temperature probes were inserted into selected root plugs while the trays of seedlings were being placed under the insulating blanket. A total of 12 temperature probes were used to monitor both air (4) and in-plug temperatures of the single-layer (2), double-layer top (2) and double-layer bottom trays (2). Two probes were also placed in control trays, inside the container medium. An automatic recording device logged temperatures from the probes hourly during the entire storage period. Probes in the control, single-layer, and double-layer bottom trays were approximately 10 cm (4 in) above ground level, whereas probes in the double-layer top trays were approximately 45 cm (18 in) above ground level.

Six trays of the same Hodwitz stock were also placed into controlled frozen storage at the former Thunder Bay Forest Nursery (TBFN), Thunder Bay, Ontario. These trays were first placed in plastic bags inside cardboard storage boxes and then overwintered in cold storage at –2 °C (28.4 °F). This stock remained in cold storage from mid-November to mid-April.

Test Configuration—Year 2. In 1991–92, the stacking configuration was changed. Although the controls remained the same, the Styroblock 165 trays under the blanket were stacked with seedling tops closer together—the tray tops were about 10 to 15 cm (4 to 6 in) apart—but with their bottoms touching and three rows high. The "single-layer" treatment was discontinued. Instead, trays of seedlings were sealed in plastic bags inside standard nursery cardboard boxes and placed in a standard shade area for overwintering. These boxes had no overwinter protection other than that provided by normal snowfall throughout the winter.

Between 3,000 and 4,000 Styroblock trays of black spruce seedlings were placed under the blanket in the second year. The trays consisted of both Styroblock 130's and 165's, with a total of about 500,000 seedlings being stored.

The Styrofoam structure was also improved (figure 2). Sand and gravel were laid down as a base to facilitate proper drainage, along with a layer of Weedmat® to prevent the growth of weeds during the summer months and to provide a better stacking and walking surface.

Temperature probes (a total of 12) were once again placed in selected root plugs of control seedlings (1), those sealed in cardboard boxes (1) and in the top (2), middle (2), and bottom (2) layer trays under the

Figure 2—*Operational setup of Hodwitz Enterprises Styrofoam® SM cold frame for the overwinter storage of container stock.*

Styrofoam blanket. In addition, 4 probes were used to monitor outdoor and underblanket ambient air temperatures. An automatic recording device logged temperatures from the probes from late-November until seedlings were removed from storage on May 1, 1992. Temperature probes for the control, cardboard box, and bottom layer underblanket were approximately 10 cm (4 in) above ground level, while probes in the middle and top underblanket trays were approximately 45 and 80 cm (18 and 31.5 in) above ground level, respectively.

No seedlings were placed into controlled frozen storage during this second year of the test.

Seedling testing and outplanting. Bud flushing tests were conducted by the Ontario Ministry of Natural Resources' North Central

6   *Tree Planters' Notes*

Region container production monitoring staff between February and April of each year of the study. Sample seedlings were removed from under the Styrofoam blanket and from the outdoor storage area, thawed, potted into a peat-vermiculite mixture, placed in a greenhouse, and monitored for bud swell and flushing.

An outplanting trial was implemented in the spring of 1991 using stock from each of the two 1990–91 insulating blanket treatments. Eighty seedlings from each of the insulating blanket and frozen storage treatments were outplanted in the spring of 1991 in the Thunder Bay Nursery outplant trial site near Raith, Ontario. Seedlings from each treatment were planted in four replicates of 20 seedlings each at 2-m (6.5-ft) spacing. The seedlings were measured in the fall of both 1991 and 1993 for total height, height increment (CAI), and root collar diameter (RCD). None of the control seedlings were outplanted. The balance of the (1991) experimentally stored seedlings were operationally planted by Canadian Pacific Forest Products (now Avenor), Thunder Bay, Ontario.

No seedlings from the 1991–92 overwintering study were experimentally outplanted in the spring of 1992. This is unfortunate, as apparently these seedlings overwintered better than those from the first year of the test. All of the overwintered seedlings were planted by Canadian Pacific Forest Products as part of their normal reforestation program.

Results

Overwintering—1990–91. Temperature monitoring showed that stock under the insulating blanket maintained fairly constant and warmer temperatures than control stock (figure 3). Minimum temperatures fluctuated less than 1 °C (33.8 °F) under the blanket in late fall, while the control fluctuated (sometimes daily) by 3 °C (37.4 °F) before sufficient snow had fallen to help in the insulation process. The seedlings were placed into storage in mid- to late November and weather records show that 14 cm (5.5 in) of snow fell on December 12th and another 10 cm (4 in) on December 20th. Temperatures inside the structure then hovered at 0 °C (32 °F) throughout the balance of the winter months, while the control

Table 1—*Monthly snowfall (cm) amounts for Thunder Bay, Ontario, during the winters of 1990–91 and 1991–92 and the 30-year average for comparison*

Month	1990–91		1991–92		30-year average	
	cm	in	cm	in	cm	in
December	50.4	19.7	53.6	20.9	46.2	18.0
January	44.0	17.2	24.8	9.7	48.4	18.9
February	15.6	6.1	28.4	11.1	30.7	12.0
March	23.6	9.2	2.4	0.9	34.2	13.3
Totals	133.6	52.1	109.2	42.6	159.5	62.2

continued to vary by up to 5 °C. Ambient temperatures for the overwintering period ranged from 0 to –35.5 °C (32 to –31 °F) (figure 3). Snowfall for the storage period can be seen in table 1.

Flushing tests conducted throughout the winter months showed that seedlings from all treatments were normal and healthy. After thawing for 6 to 13 days, the potted seedlings took from 4 to 9 days to reach full bud swell. All seedlings were fully flushed in another 4 to 5 days.

The stock monitoring staff noted that all seedlings were healthy except for those underblanket seedlings in tray plugs closest to the ground. These were prone to considerable damage, believed to be the result of warmer temperatures and higher humidity at ground level. This problem was also experienced at Jellien Nursery in Armstrong, Ontario, where a similar test was conducted with jack pine (*Pinus banksiana* Lamb.) (Neill 1991, personal communication). There are no temperature records for the Armstrong trial.

Overwintering—1991–92. Minimum temperatures recorded in the second season of this trial were considerably different from the first year. In the first year, the temperature in the control fluctuated throughout the winter, while in the second year the control remained more stable (as ambient temperatures were more moderate and snow cover more consistent) until early March, when lack of snow cover and plunging ambient temperatures allowed temperatures in the root plugs to drop considerably and rapidly. Underblanket root plug temperatures fluctuated between –1 and –5 °C (30.2 and 23°F) during this second winter (figure 4) while ambient temperatures ranged from 0 to –29°C (32 to –20.2°F).

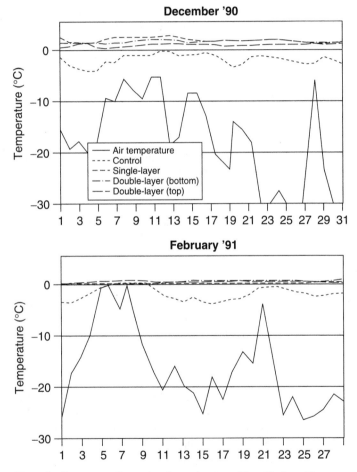

Figure 3—*Temperature fluctuations for outdoor stored Styroblock seedlings, compared to those under a Styrofoam® SM insulating blanket through the winter of 1990–91.*

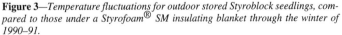

One flushing test involving the stored stock was conducted during the winter of 1991–92. Seedlings were removed from both the Styrofoam and outdoor storage areas, thawed, potted, and placed in a greenhouse. The underblanket stock all had good color with no

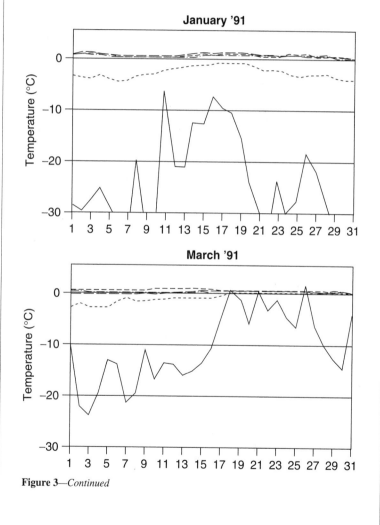

Figure 3—*Continued*

10   *Tree Planters' Notes*

damage, while some of the outdoor stock had dead terminal buds and desiccated tops.

Outplanting results. Third-year outplanting results from the black spruce seedlings overwintered under the Styrofoam® SM blanket and the

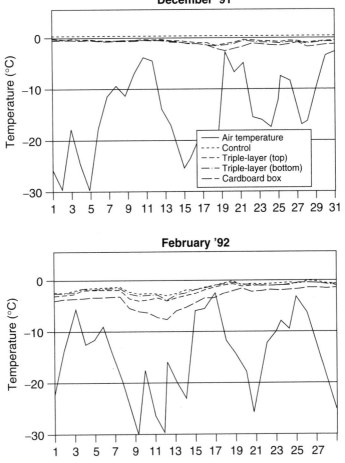

Figure 4—*Temperature fluctuations for outdoor stored Styroblock seedlings, compared to those under a Styrofoam® SM insulating blanket through the winter of 1991–92.*

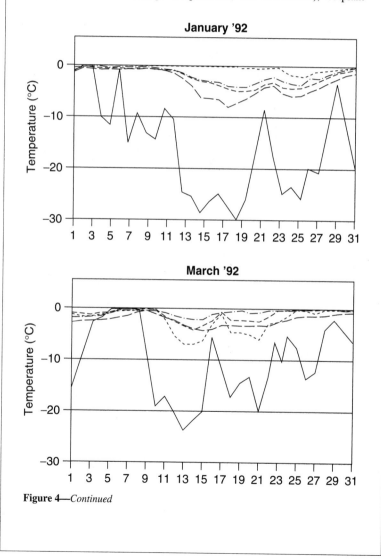

same stock overwintered in conventional cold storage at the TBFN were nearly identical (figure 5). Even though the under blanket seedlings outplanted in this test came from the first overwintering period (which had less than ideal storage temperatures, see Discussion), outplant

January '92

March '92

Figure 4—*Continued*

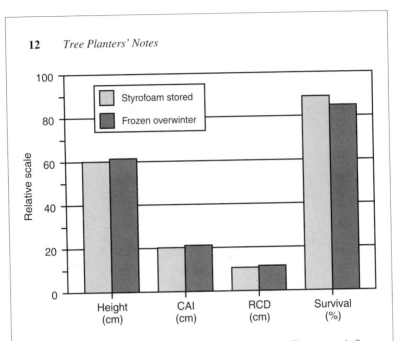

Figure 5—*Third-year outplanting results for black spruce seedlings grown in Styroblock 165's at Hodwitz Enterprises and overwintered under a Styrofoam® SM insulation blanket and in a conventional cold storage unit at the former Thunder Bay Forest Nursery. CAI = height increment, RCD = root collar diameter.*

performance of all measured parameters (height, CAI, RCD, and survival) showed no differences from that of the frozen stored stock.

Discussion

During the first year of the trial, temperatures under the blanket were observed to hover around –0 °C (32 °F). This is too warm for seedling storage (Hocking and Nyland 1971, Zalasky 1986, Odlum 1992) and can promote mold growth. It resulted in excessive damage to seedlings touching the ground, where both temperature and humidity were higher.

Changing the structure of the storage area under the blanket in the second year, by stacking the trays three high (figure 2), changed the overwintering storage temperatures and provided a better balance between cold outside air temperatures and ground heat. The storage temperatures, although somewhat variable, ranged from –1 to –5 °C (30.2 to 23 °F). This range is closer to ideal for overwintering stock, and the stock came out of storage in the spring of 1992 looking extremely healthy (Duckett 1992, personal communication).

The reduced desiccation, reduced temperature fluctuation, and reduced seedling mortality observed in this test with the use of the Styrofoam structure mirror results of a similar trial in Alberta (Matwiend).

Constructing the final overwintering storage unit cost about Can$8,000.00, with minimal maintenance costs (Hodwitz 1992, personal communication). If storage unit construction costs are depreciated over 5 years, then overwinter storage costs are about Can$6.30 per thousand. This compares to 1991 capital and operating expenses of Can$26.50 per thousand for freezer storage (Aidelbaum 1993).

Over the winter of 1992–93, Hodwitz Enterprises stored about 1.2 to 1.3 million seedlings in the unit. Loading the unit took 10 nursery workers 3 days. Unloading took a similar amount of time.

The unit should remain sealed throughout the winter, but timing of the unloading of the unit in the spring is critical. Temperature probes should be placed inside the Styrofoam blanket unit (during the loading process in the fall) to monitor underblanket temperatures in the spring. As soon as underblanket temperatures rise and remain above freezing, the unit must be opened and unloaded to prevent the seedlings from overheating.

Conclusions

Outplanting performance did not differ between the underblanket and the frozen storage treatments. However, with potential cost savings of about Can$20.00 per thousand over freezer storage, the storing of Styroblock seedlings under a Styrofoam® SM blanket seems to be a practical and economical alternative to freezer storage.

Outdoor storage of seedlings, which costs practically nothing, will continue to be used by nurseries. But this savings in overwinter storage must be weighed against the potential losses that can occur (i.e., annual losses of outdoor overwintered container stock can range between 1 and 3 million in northern Ontario, depending upon weather and seedling preconditioning (Duckett 1992, personal communication)). However, when selecting a system for overwinter storage of seedlings, numerous factors must be taken into account. These include elements of stock handling such as extraction, packaging, grading, transportation, field storage, and timing of flushing/planting.

Literature Cited

Aidelbaum AS. 1993. Frozen storage of black spruce containerized tree seedlings in northern Ontario. North Gro Development, Ltd. NorFund Project #10996. 99 p.

Gouin FR. 1977. Microfoam thermo-blanket system passes test for overwintering container plants. American Nurseryman 146(6): 11–117.

Green JL, Fuchigami LH. 1985. Protecting container-grown plants during the winter months. Ornamentals Northwestern Newsletter 9(2):10–23.

Hocking D, Nyland RD. 1971. Cold storage of coniferous seedlings: a review. Res. Rep. 6. Applied Forest Research Institute. 70 p.

Matwie L. (No date). Overwintering in insulated coldframes improves seedling survival. Hinton, AB: Weldwood of Canada, Limited. Unpublished Report.

McNiel RE, Duncan GA. 1983. Foam top poly for overwintering container stock. American Nurseryman 158(11):94–101.

Odlum KD. 1992. Frozen storage of container seedlings in Ontario. Toronto: Ontario Ministry of Natural Resources, Nursery and Seed Section. Unpublished Report. 5 p.

Ontario Ministry of Natural Resources. 1987. Report on the 1987 losses of container stock in the Thunder Bay area. Toronto. 6 p.

Zalasky H. 1986. Effects of conditioning and storage on containerized conifer seedlings. For. Mgmt. Note. 34. Canadian Forestry Service, Northern Forestry Centre. 8 p.

Index

Note: Pages containing illustrations are printed in boldface.